A Functional Analysis Framework for Modeling, Estimation and Control in Science and Engineering

A Functional Analysis Framework for Modeling, Estimation and Control in Science and Engineering

H. T. Banks

North Carolina State University
Raleigh, USA

CRC Press
Taylor & Francis Group
Boca Raton London New York

CRC Press is an imprint of the
Taylor & Francis Group, an **informa** business

A CHAPMAN & HALL BOOK

CRC Press
Taylor & Francis Group
6000 Broken Sound Parkway NW, Suite 300
Boca Raton, FL 33487-2742

First issued in paperback 2018

ISBN-13: 978-1-4398-8083-8 (hbk)
ISBN-13: 978-1-138-37463-8 (pbk)

Library of Congress Cataloging-in-Publication Data

Banks, H. Thomas.
    A functional analysis framework for modeling, estimation, and control in science and engineering / H.T. Banks.
        p. cm.
    Includes bibliographical references and index.
    ISBN 978-1-4398-8083-8 (hardback)
    1. Functional analysis. 2. Science--Mathematical models. 3. Engineering--Mathematical models. I. Title.

QA321.B36 2012
515'.7--dc23   2012009388

Contents

PREFACE

I have taught different aspects of applied functional analysis somewhat periodically since teaching my first graduate course in the Division
of Applied Mathematics at Brown University in 1968. The material in
this presentation is the result of lecture notes from courses in applied
functional analysis that I have given at the University of Southern California and North Carolina State University beginning in the late 1980s
and continuing to the present. The choices of topics covered here are
not comprehensive, and maybe even border on the eclectic in some
parts. They represent personal choices of interest to me, and represent a body of material that I want my current students and postdocs
to know. There are some standard topics, results and examples from
a typical functional analysis presentation which I hope many readers
may have seen in more traditional first courses in analysis. The notes
also reflect my own interests in results that are not readily available
in functional analysis books/lecture notes. As such these notes do not
constitute suitable material for a first course in functional analysis.

Thus some nontrivial background is desired: a first or elementary
course in analysis/functional analysis with introductory topics including linear spaces (Banach and Hilbert), linear operators, usual example
spaces $L_p(a, b)$, $C(a, b)$, etc., such as found in early chapters of classical standard texts such as [GP, LiuSob, Luen69] or the more recent
introductory text [Ped].

In these lectures, we shall present functional analysis as a basis of
modern partial and delay differential equation techniques. This is in
contrast to classical PDE techniques such as separation of variables,
Fourier transforms, Sturm-Liouville problems, etc. It is also somewhat
different from the emphasis in usual functional analysis courses where
one learns functional analysis as a subdiscipline in its own right. Here
we treat functional analysis as a **tool** to be used in understanding and
treating distributed parameter systems. We shall also motivate our
discussions with numerous application examples from biology, electromagnetics, and materials/mechanics.

Our main thrust consists of those parts and topics of functional
analysis that are fundamental to rigorous discussions of partial differential equations and delay systems (both infinite dimensional state
space systems) as they arise in diverse applications and in control and
estimation. As in standard functional analysis treatments, we will give

some proofs when we feel the urge or when we feel the arguments will have some particular instructive value. But often we simply state the results and refer readers to one of the many fine standard functional analysis texts currently available.

The presentation contains a diverse selection of applications: some are classical but some are more recent to illustrate the role that functional analysis (an old subject) continues to play in the rigorous formulation of modern applied areas. These applications are by no means inclusive. Moreover, we offer application examples far from uniform in the details given; some are more common (thermal diffusion, transport in tissue, beam vibration) and for these we offer little detail; others that are not traditional in partial differential equations, delay systems or not in the traditional functional analysis literature are given in much more detail. For example, in discussing HIV models, uncertainty in non-cooperative games, population models with aggregate data and/or dynamics, electromagnetics in materials, delay systems and partial differential equations in control and inverse problems, we have tried to err on the side of added detail. In some cases we also discuss computational aspects of the problems/applications because ultimately (again a personal belief) most serious applications result in significant computational (finite elements, spectral methods) questions.

I owe great thanks to my former and current students, postdocs, and colleagues for questions, feedback, and much proofreading. Special thanks are due to Zack Kenz and Clay Thompson for very careful readings and suggested corrections/revisions of final versions of the book. However, any errors in judgment, content, grammar, or typos in the material presented are mine!!

HTB

Acknowledgments

A substantial part of the discussions in this book concerns applications of Functional Analysis to research topics that I, my postdoctoral associates and graduate students have investigated over the past several decades. We (especially young members of our research group) have been generously supported by research grants and fellowships from US federal funding agencies including AFOSR, DARPA, NIH, NSF, DED and DOE. For this support and encouragement we are all most grateful.

1 Introduction to Functional Analysis in Applications

Before presenting a survey of preliminary but necessary foundation topics, we give one of several examples that will be used throughout our presentation, often in a motivating and illustrative spirit.

1.1 Example 1: The Heat Equation

We begin with a classical equation from thermal as well as substance transport which is based on random movement transport by Brownian mechanisms in particles or individuals. We describe this in terms of heat flow which is ubiquitous in engineering modeling [BBi2, BBCFUVW, BCCW, BCCW2, BDGJ, BDCI, BKo1, BKo2, BKoW, BT, KLB, CJ, IncropDewitt]. The basic equation for heat flow in a rod (a derivation from first principles can be found in [BT, Chapter 5]) is

$$\frac{\partial y}{\partial t}(t,\xi) = \frac{\partial}{\partial \xi}(D(\xi)\frac{\partial y}{\partial \xi}(t,\xi)) + f(t,\xi) \qquad 0 < \xi < l, \quad t > 0 \text{ (1.1)}$$

where $y(t,\xi)$ denotes the temperature in the rod at time t and position ξ and $f(t,\xi)$ is the input from a source, e.g., heat lamp, laser, etc. Example <u>Boundary Conditions (BC)</u> are

$\xi = 0:$ $y(t,0) = 0$ Dirichlet BC: This indicates temperature is held constant;

$\xi = l:$ $D(l)\frac{\partial y}{\partial \xi}(t,l) = 0$ Neumann BC: This indicates an insulated end and results in heat flux being zero.

Here $j(t,\xi) = D(\xi)\frac{\partial y}{\partial \xi}(t,\xi)$ is called the <u>heat flux</u>.

The required <u>Initial Conditions (IC)</u> are

$$y(0,\xi) = \Phi(\xi) \quad 0 < \xi < l.$$

Here Φ denotes the initial temperature distribution in the rod. The chosen boundary conditions (one constant temperature end, one insulated end) are only one combination of boundary conditions that arise in practical applications. The Dirichlet boundary conditions are called *essential* boundary conditions and must be imposed on elements in our solution spaces, while the Neumann conditions are *natural* boundary conditions and as we shall see they need not be imposed on the solution elements in weak or variational formulations of our system (see Chapter 11).

This example is typical, albeit one of the simplest, of the type of system dynamics that are of interest in our discussions in this book. Along with delay differential equations, such partial differential equation systems are rather ubiquitous in the infinite dimensional systems encountered in many practical applications in science and engineering. Our goal here is to present a functional analytic framework in which to treat modeling, parameter estimation or inverse problems and control (including feedback control) for these systems. In the research literature one finds a core of methodology for *finite dimensional* versions of these problems if one can write the underlying system dynamics in terms of a linear system

$$\begin{cases} \dot{x}(t) & = & Ax(t) + F(t) \\ x(0) & = & x_0. \end{cases} \tag{1.2}$$

Much of the partial differential and delay differential methodology developed for estimation and control theory over the past 30 years has been in the context of *infinite dimensional* systems of the form (1.2) written in an appropriate Hilbert or Banach space. Abstract representations and approximations leading to computationally feasible methods of such systems are a major focus of this book. These can be pursued in several related ways including abstract semigroup representations and/or weak/variational form representations, both of which we treat in this book.

Transforming the Initial Boundary Value Problem

If we can transform the initial boundary value problem (IBVP) for equation (1.1) into something of the form (1.2), then conceptually it could be an easier conceptual problem to treat. That is, formally it then looks like an ordinary differential equation problem for which e^{At} is a solution operator.

To rigorously make this transformation and develop a corresponding conceptual framework, we need to undertake several tasks:

1. Find a space X of functions and an operator $A : \mathcal{D}(A) \subset X \to X$ such that the IBVP can be written in the form of equation (1.2) in the space X. We may find $X = L_2(0, l)$, $C(0, l)$, $H^1(0, 1)$ or some other appropriate space (all of these are defined precisely below). But we should expect this to vary from one example to the next.

    Whatever the space, we want a solution $x(t) = y(t, \cdot)$ to (1.2), but in what mathematical sense - mild, weak, strong, classical (all terms that will be explained in detail in this treatment)? This will answer questions about regularity (i.e., smoothness) of solutions as well as suggest appropriate approximation schemes. But further questions are obvious:

2. We also want solution operators or "semigroups" $T(t)$ which play the role of e^{At}, i.e., behave like e^{At} which is written $T(t) \sim e^{At}$. But what does " e^{At} " mean in this case? If $\dot{x}(t) = Ax(t) + F(t)$, $x(0) = x_0$, is a vector system in $\mathbb{R}^n$ so that $A \in \mathbb{R}^n \times \mathbb{R}^n$ is a constant matrix, then

$$e^{At} \equiv \sum_{n=0}^{\infty} \frac{(At)^n}{n!} = I + At + \frac{A^2 t^2}{2!} + \dots$$

    where the series has nice convergence properties. In this case, by the "variation of constants" or "variation of parameters" representations, we can then write

$$x(t) = e^{At} x_0 + \int_0^t e^{A(t-s)} F(s) ds.$$

    We want a similar variation of constants representation in X with operators $T(t) \in \mathcal{L}(X) = \mathcal{L}(X, X)$ such that $T(t) \sim e^{At}$, so that

$$x(t) = T(t) x_0 + \int_0^t T(t-s) F(s) ds \qquad (1.3)$$

    holds, represents the PDE solutions in some appropriate sense, and can be used for qualitative (stability, asymptotic behavior, control) and quantitative (approximation and numerics) analyses.

    We note that e^{At} has what are known as *semigroup* properties: $e^{A(t+s)} = e^{At} e^{As}$, $e^{A0} = I$, so that solutions $x(t) = e^{At} x_0$ of the homogeneous equation ($F = 0$ in (1.2)) also have this translation invariance property characteristic of autonomous dynamical systems. It is precisely this property that we abstract in a theory of linear dynamic operators $T(t)$ in the context of linear semigroups in abstract Hilbert and Banach spaces. Before we present this theory in the next chapter, we summarize some needed notation and also present several other examples arising in applications that will be used throughout this book to illustrate ideas and results.

1.2   Some Preliminaries: Hilbert, Banach, and Other Spaces Useful in Operator Theory

Before proceeding, we recall some standard definitions and results from analysis and elementary functional analysis found in numerous references such as [Adams, CH, D, DS, E, HeSt, K, Kreyszig, NS, Ped, RN, Ru1, Ru2, Sh, W, Y].

Normed Linear Spaces and Linear Operators

A *normed linear space* X over a scalar field $\mathbb{F}$ (in all our examples $\mathbb{F}$ is either the real $\mathbb{R}^1$ or complex $\mathbb{C}$ number system) is a *linear* space (closed under addition of elements and multiplication by scalars with a *norm* $|\cdot|$ (in many references the notation $\|\cdot\|$ is also used) which is a map from the space to the nonnegative numbers satisfying (i) $|cx| = |c||x|$, (positive homogeneity or positive scalability), (ii) $|x_1 + x_2| \le |x_1| + |x_2|$ (triangle inequality or subadditivity), and (iii) If $|x| = 0$, then x is the zero vector (separates points). In a normed linear space any sequence $\{x_n\}$ that satisfies $|x_n - x_m| \to 0$ as $n, m \to \infty$ is called a *Cauchy* sequence. A normed linear space is called a *complete normed linear space* or a *Banach space* if for any Cauchy sequence $\{x_n\}$ in X there exists $x \in X$ such that $\lim_{n \to \infty} x_n = x$.

A *Hilbert space* is a Banach space with scalar field $\mathbb{F}$ which also has an *inner product* $\langle \cdot, \cdot \rangle : X \times X \to \mathbb{F}$ satisfying (i) $\langle x, y \rangle = \overline{\langle y, x \rangle}$ (conjugate symmetry which in case $\mathbb{F} = \mathbb{R}^1$ yields symmetry), (ii) $\langle cx_1, x_2 \rangle = c\langle x_1, x_2 \rangle, \langle x_1 + x_2, x_3 \rangle = \langle x_1, x_3 \rangle + \langle x_2, x_3 \rangle$ (linearity in the first argument), (iii) $\langle x, x \rangle \ge 0$ with equality holding only for $x = 0$ (positive-definiteness). We observe that conjugate symmetry implies that $\langle x, x \rangle$ is real for all x, since one has $\langle x, x \rangle = \overline{\langle x, x \rangle}$.

A *bounded linear operator* is a linear transformation T between normed vector spaces X and Y for which the ratio of the norm of $T(x)$ to that of x is bounded by the same number, over all non-zero vectors x in X. In other words, there exists some $M > 0$ such that $|Tx|_Y \leq M|x|_X$ for all x in X. The smallest such M is called the operator norm $|T|$ of T (see below).

A bounded linear operator is generally not a bounded function; the latter would require that the $|T(x)|$ be bounded for all x, which is possible only if $Y = \{0\}$, i.e., Y is the zero vector space.

A standard exercise in elementary functional analysis is to argue that a linear operator is bounded if and only if it is continuous.

Let X, Y be normed linear spaces. Then

$$\mathcal{L}(X, Y) = \{T : X \to Y \mid T \text{ is bounded, linear}\}$$

is also a normed linear space with norm given by

$$|T|_{\mathcal{L}(X,Y)} = \sup_{|x|_X = 1} |Tx|_Y = \sup_{x \neq 0} \frac{|Tx|_Y}{|x|_X}.$$

Note that the notation $\mathcal{B}(X, Y)$ is also frequently used in the literature. When $Y = X$ one often writes $\mathcal{L}(X)$ or $\mathcal{B}(X)$ for $\mathcal{L}(X, X)$ or $\mathcal{B}(X, X)$. When Y is chosen as the scalar field $\mathbb{F}$ for X, we define the *linear dual space* of X by $X^D \equiv \mathcal{L}(X, \mathbb{F})$, the normed linear space of all continuous linear functionals (transformations of functions with range in the scalar field) on X. While the linear dual space of a space X is sometimes denoted by X^*, we shall throughout this book reserve this notation for what is sometimes called the anti-dual or conjugate dual space.

We define *conjugate linear functionals* or *anti-linear functionals* as functionals $T : X \to \mathbb{F}$ satisfying $T(\alpha x + \beta y) = \bar{\alpha} T(x) + \bar{\beta} T(y)$ for all $x, y \in X$ and $\alpha, \beta \in \mathbb{F}$. Note that if $\mathbb{F} = \mathbb{R}^1$ then conjugate linear and linear are the same. For a normed linear space X we define X^*, called the *conjugate dual space* or *adjoint space*, as the linear space of continuous conjugate linear or anti-linear functionals on X. It is readily established that X^* is a Banach space whenever X is a normed linear space and the scalar field is complete. The elements of X^* will be denoted by x^* and the norm $|\cdot|_*$ in X^* is given by $|x^*|_* = \sup_{|x| \leq 1} |x^*(x)|$.

In this book we shall have little use for X^D but will heavily use X^* and simply call it the *dual space* throughout. The heavy use of this adjoint or dual space X^* is not unusual (see [K, p.11], [W, p.165], [Sh, p.5,9]) when treating partial differential equations in the context of sesquilinear or semi-linear forms as we shall do in this book.

Exercise: Show that $\mathcal{L}(X, Y)$ is complete if Y is complete.

We shall also need throughout our examples some specific spaces found in the literature [Adams, DS]. For the usual Lebesgue and Sobolev spaces we use the notation $L_p(a, b)$ or $L_p(a, b; Y)$ and $H^p(a, b)$ (typically $p = 1$ or $p = 2$ for our discussions in this book). For the space of continuous functions with supremum norm we use $C(a, b)$. In general, $W^{k,p}(a, b) = \{\varphi \in L_p | \varphi', \varphi'', \ldots, \varphi^{(k)} \in L_p\}$. If $p = 2$, we write $W^{k,2}(a, b) = H^k(a, b)$, $k = 1, 2, \ldots$ Finally, $AC(a, b) = \{\varphi \in L_2(a, b) | \varphi$ is absolutely continuous (AC) on $[a, b]\}$.

Differentiation in Normed Linear Spaces

Again these are standard definitions that can be found in numerous references including [HP].

Suppose $f : X \to Y$ is a (nonlinear) transformation (mapping).

Definition 1.1 If $\lim_{\epsilon \to 0^+} \frac{f(x_0 + \epsilon z) - f(x_0)}{\epsilon}$ exists, we say f has a *directional* or *Gateaux* derivative at x_0 in the direction z. This is denoted $\delta f(x_0; z)$, and is called the *Gateaux differential* at x_0 in the direction z.

If the limit exists for any direction z, we say f is Gateaux differentiable at x_0 and $z \to \delta f(x_0; z)$ is the *Gateaux derivative*. Note that $z \to \delta f(x_0; z)$ is not necessarily linear, however it is homogeneous of degree one (i.e., $\delta f(x_0; \lambda z) = \lambda \delta f(x_0; z)$ for scalars λ). Moreover, it need not be continuous in z.

The following definition of $o(|z|)$ will be useful in defining the Fréchet derivative:

Definition 1.2 The function $g(z)$ is $o(|z|)$ if $\frac{|g(z)|}{|z|} \to 0$ as $z \to 0$.

Definition 1.3 If there exists $df(x_0; \cdot) \in \mathcal{L}(X, Y)$ such that $|f(x_0 + z) - f(x_0) - df(x_0; z)|_Y = o(|z|_X)$, then $df(x_0; \cdot)$ is the *Fréchet derivative* of f at x_0. Equivalently, $\lim_{\epsilon \to 0} \frac{f(x_0 + \epsilon z) - f(x_0)}{\epsilon} = df(x_0; z)$ for every $z \in X$ with $z \to df(x_0; z) \in \mathcal{L}(X, Y)$. We write $df(x_0; z) = f'(x_0)z$.

Summary Results

(These also could be taken as exercises by industrious readers!)

1. If f has a Fréchet derivative, it is unique.

2. If f has a Fréchet derivative at x_0, then it also has a Gateaux derivative at x_0 and they are the same.

3. Let D be an open subset of X. If $f : D \subset X \to Y$ has a Fréchet derivative at $x_0 \in D$, then f is continuous at x_0.

Examples

1. $X = \mathbb{R}^n$, $Y = \mathbb{R}^m$, and $f : X \to Y$ such that each component of f has partial derivatives at x_0 in the form $\frac{\partial f_i}{\partial x_j}(x_0)$. Then the Fréchet derivative can be represented as a matrix of partials

$$f'(x_0) = \left(\frac{\partial f_i}{\partial x_j}(x_0) \right)$$

and hence

$$df(x_0; z) = \left(\frac{\partial f_i}{\partial x_j}(x_0) \right) z.$$

2. Now we consider the case where $X = \mathbb{R}^n$ and $Y = \mathbb{R}^1$. Then

$$df(x_0; z) = \nabla f(x_0) \cdot z.$$

3. We also consider the case where $X = Y = \mathbb{R}^1$. Here,

$$df(x_0; z) = f'(x_0)z.$$

Normally $z = 1$ because that is the *only* direction in $\mathbb{R}^1$. Then $f'(x_0)$ is called the derivative, but in actuality $z \to f'(x_0)z$ is the derivative. Note that $f'(x_0) \in \mathcal{L}(\mathbb{R}^1, \mathbb{R}^1)$, but the elements of $\mathcal{L}(\mathbb{R}^1, \mathbb{R}^1)$ are just numbers.

Homework Exercises

- Ex. 1 : Consider $f : \mathbb{R}^2 \to \mathbb{R}^1$.

$$f(x_1, x_2) = \begin{cases} \frac{x_1 x_2^2}{x_1^2 + x_2^4}, & \text{if } x \neq 0 \\ 0, & \text{if } x = 0 \end{cases}$$

Show that f is Gateaux differentiable at $x = (x_1, x_2) = 0$, but f is not continuous at $x = 0$, (and hence cannot be Fréchet differentiable).

1.3   Return to Example 1: Heat Equation

To begin the process of writing the system of Example 1 in the form of equation (1.2), take $X = L_2(0,l)$ and define

$$\mathcal{D}(A) = \{\varphi \in H^2(0,l) | \varphi(0) = 0, \varphi'(l) = 0\}$$

to be the domain of A. (Assume D is smooth for now, e.g., D is at least H^1.) Then we can define $A : \mathcal{D}(A) \subset X \to X$ by

$$A\varphi(\xi) = (D(\xi)\varphi'(\xi))'. \tag{1.4}$$

Homework Exercises

- Ex. 2 : Show that $A : \mathcal{D}(A) \subset X \to X$ of the heat example (Example 1) is <u>not</u> a bounded operator from $X \to X$.

- Ex. 3 : Give an example when X is a Hilbert space, but $\mathcal{L}(X)$ is not a Hilbert space.

- Ex. 4 : Consider $H^1(a,b) = W^{1,2}(a,b), W^{1,1}(a,b)$ and $AC(a,b)$. Find the relationships for each pair of spaces in terms of $\subset$. That is, establish $X \subset Y$ or $X \subsetneq Y$, etc.

Before beginning our presentation on linear semigroups, we give a number of other motivating examples to be used in the sequel along with the heat equation example above.

1.4   Example 2: General Transport Equation

Problems in Which the Transport Equation Is Used

After the heat or diffusion equation of Example 1, probably the most widely encountered system in applications is the so-called transport equation which embodies both diffusive as well as convective movement of species, particles, or individuals. Included in a diverse range of applications are insect dispersion involving random movement (Brownian motion or "diffusion") as well as some type of directed (convective, advective) movement [BKa, BK, BDK, BKL, BKZ] that often is a function of external stimulatory signals. These equations also arise (as backward Kolomogorov equations) in population growth models in which some randomness or stochasticity is present [Allen, Gard, GRD-FP, GRD-FP2, Murray, Okubo]. They are also a basis for models in many "flow" problems (convective/diffusive transport) in ecology and physiology (e.g., see the "cat brain problem" [BKa, BK] which involves in vivo labeled transport to investigate pure diffusion vs. diffusion plus convected transport in brain tissue in anesthetized cats). Such equations also arise in distributed models for mixing mechanisms in lake and sea sediment cores [BR1, BR2] as well as in other more recent label transport and decay models [BSTBRSM, BST2, BCDST]. Transport equations with source/sink, advection, and diffusion terms provide analogs for modeling different components of continent-scale population dynamics in Holocene changes in beech pollen percentages [DBW] and suggest that changes in beech populations reflect spatial shifts in the multifactorial environmental conditions that are most favorable for the growth of beech trees. The basic ideas and concepts underlying these myriad applications are easy to describe in the setting of one-dimensional equations.

Description of Various Fluxes

In these models there are several characteristic components that contribute to the "flux" or movement. These can be described by terms such as (here the quantity $y(t, \xi)$ is the population or amount of a substance at location ξ at time t in the habitat)

- $j_1(t, \xi)$ is the flux due to dispersion (random foraging or moments; random molecular collisions) where

$$j_1(t, \xi) = D(\xi) \frac{\partial y}{\partial \xi}(t, \xi)$$

- $j_2(t, \xi)$ is the advective/convective/directed bulk movement where
$$j_2(t, \xi) = \nu(\xi) y(t, \xi)$$

- $j_3(t, \xi)$ is the net loss due to birth/death, immigration/emigration, label decay, where

$$j_3(t, \xi) = \mu(\xi) y(t, \xi)$$

General Transport Equation for a 1-Dimensional Habitat

The general transport equation is readily derived from mass balance first principles (see [BT, Chapter 4]) and is given by

$$\frac{\partial y}{\partial t}(t,\xi) + \frac{\partial}{\partial \xi}(\nu(\xi)y(t,\xi)) = \frac{\partial}{\partial \xi}(D(\xi)\frac{\partial y}{\partial \xi}(t,\xi)) - \mu(\xi)y(t,\xi),$$
$$0 < \xi < l, \quad t > 0 \qquad (1.5)$$

B.C.

$\xi = 0:$ $y(t,0) = 0$ (essential)

$\xi = l:$ $\left[D(\xi)\frac{\partial y}{\partial \xi}(t,\xi) - \nu(\xi)y(t,\xi)\right]_{\xi=l} = 0$ (natural)

I.C., $y(0,\xi) = \Phi(\xi)$

In preparation to later write this example in operator form (1.2), we let $X = L_2(0,l)$ be the usual complex Hilbert space of square integrable functions, and the domain of A be defined by $\mathcal{D}(A) = \{\varphi \in H^2(0,l)|\varphi(0) = 0, [D\varphi' - \nu\varphi](l) = 0\}$ where

$$A\varphi(\xi) = (D(\xi)\varphi'(\xi))' - (\nu(\xi)\varphi(\xi))' - \mu\varphi(\xi). \qquad (1.6)$$

Note: We will assume D and ν are smooth for now.

1.5   Example 3: Delay Systems–Insect/Insecticide Models

Delay systems have been of interest for the past 70 or more years, arising in applications ranging from aerospace engineering to biology (biochemical pathways, etc.), population models, ecology, HIV and other disease progression models to viscoelastic and smart hysteretic materials, as well as network models [BBH, BBJ, BBPP, BKurW1, BKurW2, BRS, Hutch, KP, Ma, Vis, Warga72, Wright]. We describe here one arising in insect/insecticide investigations [BBJ].

    Mathematical models that are suitable for field data with mixed populations should consider reproductive effects and should also account for multiple generations, containing neonates (juveniles) and adults and their interconnectedness. This suggests the need at the minimum for a coupled system of equations describing two separate age classes. Additionally, due to individual differences within the insect population, it is biologically unrealistic to assume that all neonate aphids born on the same day reach the adult age class at the same time. In fact, the age at which the insects reach adulthood varies from as few as five to as many as seven days. Hence one must include a term in any model to account for this variability, leading one to develop a coupled delay differential equation model for the insect population dynamics. We consider the delay between birth and adulthood for neonate pea aphids and present a first mathematical model that treats this delay as a random variable. For a careful derivation of models with similar structure in HIV progression at the cellular level, see [BBH].

Let $a(t)$ and $n(t)$ denote the number of adults and neonates, respectively, in the population at time t. We lump the mortality due to insecticide into one parameter p_a for the adults, p_n for the neonates, and denote by d_a and d_n the background or natural mortalities for adults and neonates, respectively. We let b be the rate at which neonates are born into the population.

We suppose that there is a time delay for maturation of a neonate to the adult life stage. We further assume that this time delay varies across the insect population according to a probability distribution $P(\tau)$ for $\tau \in [-T_n, 0]$ with corresponding density $k(\tau) = \frac{dP(\tau)}{d\tau}$. Here we tacitly assume an upper bound on T_n for the maturation period of neonates into adults. Thus, we have that $k(\tau)$, $\tau < 0$, is the probability per unit time that a neonate who has been in the population $-\tau$ time units becomes an adult. Then the rate at which such neonates become adults is $n(t+\tau)k(\tau)$. Summing over all such τ's, we obtain that the rate at which neonates become adults is $\int_{-T_n}^{0} n(t+\tau)k(\tau)d\tau$. Using the biological knowledge that the maturation process varies between five and seven days (i.e., k vanishes outside $[-7, -5]$), we obtain the functional differential equation (FDE) system

$$
\begin{aligned}
\frac{da}{dt}(t) &= \int_{-7}^{-5} n(t+\tau)k(\tau)d\tau - (d_a + p_a)\,a(t) \\
\frac{dn}{dt}(t) &= ba(t) - (d_n + p_n)\,n(t) - \int_{-7}^{-5} n(t+\tau)k(\tau)d\tau \quad\quad (1.7) \\
a(\theta) &= \Phi(\theta), \quad n(\theta) = \Psi(\theta), \quad \theta \in [-7, 0) \\
a(0) &= a^0, \quad\quad n(0) = n^0,
\end{aligned}
$$

where k is now a probability density kernel ($\int_{-7}^{-5} k(\tau)d\tau = 1$) which we have assumed has the property $k(\tau) \geq 0$ for $\tau \in [-7, -5]$ and $k(\tau) = 0$

for $\tau \in [-T_n, -7) \cup (-5, 0]$.

The system of functional differential equations described in (1.7) can be written in terms of an equation of the form (1.2) [BBu1, BBu2, BKap1] as follows.

Let

$$z(t) = (a(t), \, n(t))^T$$

and

$$z_t(\tau) = z(t + \tau), \, -7 \le \tau \le 0. \tag{1.8}$$

We define the Hilbert space $X \equiv \mathbb{R}^2 \times L_2(-7, 0; \mathbb{R}^2)$ with norm (and obvious corresponding inner product)

$$|(\eta, \varphi)|_X = \left(|\eta|^2 + \int_{-7}^0 |\varphi(\theta)|^2 d\theta \right)^{1/2}, \, (\eta, \varphi) \in X,$$

and let $x(t) = (z(t), z_t) \in X$. Then our system (1.7) can be written as

$$\begin{aligned} &\frac{dz}{dt}(t) = L(z_t) \text{ for } 0 \le t \le T, \\ &(z(0), z_0) = (\eta, \varphi) \in X, \quad (\eta, \varphi) \in \mathbb{R}^2 \times L_2(-7, 0; \mathbb{R}^2), \end{aligned} \tag{1.9}$$

where $T < \infty$ and for $(\eta, \varphi) \in \mathbb{R}^2 \times L_2(-7, 0; \mathbb{R}^2)$,

$$L(\varphi) = \begin{bmatrix} -d_a - p_a & 0 \\ b & -d_n - p_n \end{bmatrix} \varphi(0) + \begin{bmatrix} 0 & 1 \\ 0 & -1 \end{bmatrix} \int_{-7}^{-5} \varphi(\tau)k(\tau)d\tau. \tag{1.10}$$

We now define a linear operator $\mathcal{A} : \mathcal{D}(\mathcal{A}) \subset X \to X$ with domain

$$\mathcal{D}(\mathcal{A}) = \{ (\varphi(0), \varphi) \in X \, | \, \varphi \in H^1(-7, 0; \mathbb{R}^2) \} \tag{1.11}$$

by

$$\mathcal{A}(\varphi(0), \varphi) = (L(\varphi), \varphi'). \tag{1.12}$$

Then the delay system (1.7) can be formulated as

$$\begin{aligned} \dot{x}(t) &= \mathcal{A}x(t) \\ x(0) &= x_0, \end{aligned} \tag{1.13}$$

where $x_0 = (\eta_0, \varphi_0) = ((a^0, n^0)^T, (\Phi, \Psi)^T)$.

As we shall see in subsequent sections of this book, FDE systems [JKH1, JKH2, JKH3] are of widespread interest for many years dating back at least to Minorsky [Minorsky].

1.6 Example 4: Probability Measure Dependent Systems — Maxwell's Equations

The equations and ideas underlying the theory of electricity and magnetism in materials (where these concepts are most important) are quite complex. They are frequently described in engineering and physics with a rigor and preciseness that is less than satisfactory for mathematicians. Therefore we will give a somewhat more detailed presentation to these models and applications than in our earlier examples for thermal and transport processes. We consider Maxwell's equations in a complex, heterogeneous material (see [BBL] and the extensive electromagnetics references therein including [Balanis, Cheng, Jackson, Elliot, RMC] for details):

$$\nabla \times E = \frac{-\partial B}{\partial t}$$
$$\nabla \times H = \frac{\partial D}{\partial t} + J$$
$$\nabla \cdot D = \rho$$
$$\nabla \cdot B = 0$$

where E is the electric field (force), H is the magnetic field (force), D is the electric flux density (also called the electric displacement), B is the magnetic flux density, and ρ is the density of charges in the medium.

To complete this system, we need constitutive (material dependent) relations:

$$D = \epsilon_0 E + P$$
$$B = \mu_0 H + M$$
$$J = J_c + J_s$$

where P is electric polarization, M is magnetization, J_c is the conduction current density, J_s is the source current density, ϵ_0 is the dielectric permittivity, and μ_0 is the magnetic permeability in free space.

General polarization (a constitutive law) is given by:

$$P(t, \bar{x}) = [\hat{g} * E](t, \bar{x}) = \int_0^t \hat{g}(t - s, \bar{x}) E(s, \bar{x}) ds$$

Here, $\hat{g}$ is the polarization susceptibility kernel, or dielectric response function (DRF).

Two Widely Used Examples of Polarization:

- **Debye Model for Polarization**
  This describes reasonably well a polar material. This is also called dipolar or orientational polarization. The DRF is defined by

$$\hat{g}(t - s, \bar{x}) = g(t - s, \bar{x}) + \epsilon_0(\epsilon_\infty - 1)\delta(t - s) \qquad (1.14)$$

where δ is the Dirac delta function and

$$g(t - s, \bar{x}) = e^{\frac{-(t-s)}{\tau}} \left(\frac{\epsilon_0(\epsilon_s - \epsilon_\infty)}{\tau} \right). \qquad (1.15)$$

This corresponds to $P = \epsilon_0(\epsilon_\infty - 1)E + \tilde{P}$ where $\tilde{P}$ satisfies

$$\dot{\tilde{P}} + \frac{1}{\tau}\tilde{P} = \epsilon_0(\epsilon_s - \epsilon_\infty)E.$$

It is important to note that P represents *macroscopic* polarization, as opposed to *microscopic polarization p* (introduced below) which describes polarization at the molecular level.

- **Lorentz Polarization**
  This is also called electronic polarization or the electronic cloud model. The DRF is again given by (1.14) but now with g defined by

$$g(t - s, \bar{x}) = \frac{\epsilon_0 \omega_p^2}{\nu_0} e^{\frac{-(t-s)}{2\tau}} \sin(\nu_0(t - s)). \qquad (1.16)$$

This corresponds to $P = \epsilon_0(\epsilon_\infty - 1)E + \tilde{P}$ where $\tilde{P}$ satisfies

$$\ddot{\tilde{P}} + \frac{1}{\tau}\dot{\tilde{P}} + \omega_0^2\tilde{P} = \epsilon_0\omega_p^2 E$$

and $\omega_p = \omega_0\sqrt{\epsilon_s - \epsilon_\infty}$ and $\nu_0 = \sqrt{\omega_0^2 - \frac{1}{4\tau^2}}$.

For complex composite materials, the standard Debye or Lorentz polarization model is not adequate, e.g., we need multiple relaxation times τ's in some kind of distribution [BBo, BG1, BG2]. As an example the multiple Debye model becomes

$$P(t, \bar{x}; F) = \int_\mathcal{T} p(t, \bar{x}; \tau)dF(\tau),$$

where $\mathcal{T}$ is a set of possible relaxation parameters τ,

$$F \in \mathcal{F}(\mathcal{T}) = \{F : \mathcal{T} \to \mathbb{R}^1 | F \text{ is a probability distribution on } \mathcal{T}\},$$

and p is the microscopic polarization given by

$$p(t, \bar{x}; \tau) = \int_0^t \hat{g}(t - s, \bar{x}; \tau) E(s, \bar{x}) ds.$$

Thus,

$$
\begin{aligned}
P(t, \bar{x}; F) &= \int_{\mathcal{T}} \int_0^t \hat{g}(t - s, \bar{x}; \tau) E(s, \bar{x}) ds dF(\tau) \\
&= \int_0^t \left[\int_{\mathcal{T}} \hat{g}(t - s, \bar{x}; \tau) dF(\tau) \right] E(s, \bar{x}) ds \\
&= \int_0^t G(t - s, \bar{x}; F) E(s, \bar{x}) ds
\end{aligned}
$$

Assuming $\rho = 0, M = 0$ (no fixed charges, nonmagnetic materials), we find this system becomes

$$
\begin{aligned}
\nabla \times E &= -\frac{\partial}{\partial t}(\mu_0 H) \\
\nabla \times H &= \frac{\partial}{\partial t}\left[\epsilon_0 E + \int_0^t G(t - s, \bar{x}; F) E(s, \bar{x}) ds \right] + J \\
\nabla \cdot D &= 0 \\
\nabla \cdot H &= 0
\end{aligned}
\tag{1.17}
$$

where $F \in \mathcal{F}(\mathcal{T})$ and $J = J_c + J_s$. Note: J_c is usually also a convolution on E although Ohm's law uses $J_c = \sigma E$ where σ is the conductivity of the material. In general, one should treat J_c as a convolution, e.g.,

$$J_c = \sigma_c * E = \int_0^t \sigma_c(t - s, \bar{x}) E(s, \bar{x}) ds.$$

For the measure dependent system (1.17), existence and uniqueness via a semigroup formulation have not been established. Comparisons of solutions via semigroups versus via weak solutions have not been done. Little has yet been done in two or three dimensions [BaBar]. For the one-dimensional case only, existence, uniqueness, and continuous dependence have been established via a weak formulation (see [BG1]). For continuous dependence of solutions on F, a metric is needed on $\mathcal{F}(\mathcal{T})$. More generally, we may need to treat other material parameters $q = (\tau, \epsilon_s, \epsilon_\infty, \sigma)$, where $q \in Q \subset \mathbb{R}^4$ and we look for $F \in \mathcal{F}(Q)$.

• **Special Case**

We can consider a physically meaningful special case of the Maxwell system in a dielectric material which has a general polarization convolution relationship. Detailed derivations given in Section 2.3 of [BBL] lead to a one-dimensional version we next present and will use for an example in our subsequent discussions. Among the assumptions are some homogeneity in the medium (in planes parallel to that of an interrogating polarized planar wave) and a polarized sheet antenna source J_s. The resulting model leads to only nontrivial E fields in the x direction, and H fields in the y direction, each depending only on t and z. Assuming Ohm's law for J_c, we find the system reduces to

$$
\begin{aligned}
\frac{\partial E}{\partial z} &= -\mu_0 \frac{\partial H}{\partial t} \\
-\frac{\partial H}{\partial z} &= \frac{\partial D}{\partial t} + \sigma E + J_s \\
D &= \epsilon_0 E + P
\end{aligned}
\tag{1.18}
$$

or in second-order form for $E(t, z)$:

$$
\mu_0 \epsilon_0 \frac{\partial^2 E}{\partial t^2} + \mu_0 \frac{\partial^2 P}{\partial t^2} + \mu_0 \sigma \frac{\partial E}{\partial t} - \frac{\partial^2 E}{\partial z^2} = -\mu_0 \frac{\partial J_s}{\partial t}
$$

or

$$
\frac{\partial^2 E}{\partial t^2} + \frac{1}{\epsilon_0} \frac{\partial^2 P}{\partial t^2} + \frac{\sigma}{\epsilon_0} \frac{\partial E}{\partial t} - c^2 \frac{\partial^2 E}{\partial z^2} = \frac{-1}{\epsilon_0} \frac{\partial J_s}{\partial t},
$$

where $c^2 = \frac{1}{\mu_0 \epsilon_0}$. Typical boundary conditions (say on $\Omega = [0, 1]$) are

$$
\left[\frac{1}{c} \frac{\partial E}{\partial t} - \frac{\partial E}{\partial z} \right]_{z=0} = 0 \qquad \text{(absorbing B.C. at } z = 0)
$$

$$
\text{and} \qquad E|_{z=1} = 0 \qquad \text{(supraconductive B.C. at } z = 1).
$$

We assume a general polarization relationship

$$
P(t, z) = \int_0^t \hat{g}(t - s, z) E(s, z) ds = \int_0^t g(t - s, z) E(s, z) ds + \hat{\beta} E(t, z),
$$

where g is absolutely continuous and $\hat{\beta}$ contains coefficients for any Dirac components of $\hat{g}$, and initial conditions

$$
E(0, z) = \Phi(z), \qquad \frac{\partial E(0, z)}{\partial t} = \Psi(z).
$$

Then it can be argued [BBL, Section 2.3] that the system becomes

$$\frac{\partial^2 E}{\partial t^2} + \gamma \frac{\partial E}{\partial t} + \beta E + \int_{-\infty}^{0} \tilde{g}(s) E(t+s) ds - c^2 \frac{\partial^2 E}{\partial z^2} = \mathcal{J}(t),$$

(for theoretical considerations and without loss of generality, we may take $\epsilon_0 = 1$, $\hat{\beta} = 0$ and assume g independent of z) where $\beta, \gamma, \mathcal{J}$ are defined as in [BBL, p. 21] and we have tacitly assumed $E(t, z) = 0$ for $t < 0$. Here $\tilde{g}$ is defined by $\tilde{g}(s) = \ddot{g}(-s)$. If we approximate the "memory" term by assuming only a finite past history is significant, we obtain the integro-partial differential equation

$$\frac{\partial^2 E}{\partial t^2} + \gamma \frac{\partial E}{\partial t} + \beta E + \int_{-r}^{0} \tilde{g}(s) E(t+s) ds - c^2 \frac{\partial^2 E}{\partial z^2} = \mathcal{J}(t). \quad (1.19)$$

As with the first three examples, this example can also be written as $\frac{dx}{dt} = Ax + F$ in an appropriate function space setting. In this direction we define an operator A in appropriate state spaces. In this example one might choose the electric field E and the magnetic field (H in (1.18)) as "natural" states along with some type of hysteresis state to account for the memory in (1.19). However, in second-order (in time) systems, it is also sometimes natural to choose the state E and velocity $\frac{\partial E}{\partial t}$ as states. We do that in this example. We first define an auxiliary variable $w(t)$ in a weighted L_2 space $W \equiv L_2(-r, 0; \tilde{g}; L_2(\Omega))$ by

$$w(t)(\theta) = E(t) - E(t+\theta), \qquad -r \leq \theta \leq 0.$$

We note that for each t and θ, $w(t)(\theta)$ is itself a function in $L_2(\Omega)$, i.e., a function of z given by $w(t)(\theta, z) = E(t, z) - E(t+\theta, z)$. For the inner product in W, we choose the weighted inner product

$$\langle \eta, w \rangle_W \equiv \int_{-r}^{0} \tilde{g}(\theta) \langle \eta(\theta), w(\theta) \rangle_{L_2(\Omega)} d\theta$$

and then (1.19) can be written as

$$\frac{\partial^2 E}{\partial t^2} + \gamma \frac{\partial E}{\partial t} + (\beta + g_{11}) E - \int_{-r}^{0} \tilde{g}(s) w(t)(s) ds - c^2 \frac{\partial^2 E}{\partial z^2} = \mathcal{J}(t), \quad (1.20)$$

where $g_{11} \equiv \int_{-r}^{0} \tilde{g}(s) ds$ and $w(t)(s) = E(t) - E(t+s)$, $-r \leq s \leq 0$. For a semigroup formulation we consider the state space

$$X = V \times H \times W = H_R^1(\Omega) \times L_2(\Omega) \times L_2(-r, 0; \tilde{g}; H)$$

(here $H^1_R(\Omega) = \{\phi \in H^1 | \phi(1) = 0\}$) with states

$$(\phi, \psi, \eta) = \left(E(t), \frac{\partial E(t)}{\partial t}, w(t) \right) = \left(E(t), \frac{\partial E(t)}{\partial t}, E(t) - E(t + \cdot) \right).$$

We note that the magnetic field H defined above is no longer a part of the model and hence no confusion should result if we use H to denote the space $L_2(\Omega)$.

To define what we will later see is an infinitesimal generator of a C_0 semigroup, we first define several component operators. Let $\hat{A} \in \mathcal{L}(V, V^*)$ be formally defined by

$$\hat{A}\phi \equiv c^2 \phi'' - (\beta + g_{11})\phi + c^2 \phi'(0)\delta_0, \tag{1.21}$$

where δ_0 is the Dirac operator $\delta_0 \psi = \psi(0)$ and the precise meaning for ϕ'' in this case for $\phi \in H^1_R(\Omega)$ will be given later (see for example (3.20) in Section 3.6.4). Here V^* is the dual space of V (i.e., the space of continuous linear functionals on V). We also define $B \in \mathcal{L}(V, V^*)$ and $\hat{K} \in \mathcal{L}(W, H)$ by

$$B\phi = -\gamma\phi - c\phi(0)\delta_0$$

and, for $\eta \in W$,

$$(\hat{K}\eta)(z) = \int_{-r}^{0} \tilde{g}(\theta)\eta(\theta, z)d\theta, \qquad z \in \Omega.$$

Moreover, we introduce the operator $C : \mathcal{D}(C) \subset W \to W$ defined on

$$\mathcal{D}(C) = \{\eta \in H^1(-r, 0; L_2(\Omega)) \mid \eta(0) = 0\}$$

by

$$C\eta(\theta) = \frac{\partial}{\partial\theta}\eta(\theta).$$

We then define the operator $\mathcal{A}$ on

$$\mathcal{D}(\mathcal{A}) = \{(\phi, \psi, \eta) \in X | \psi \in V, \eta \in \mathcal{D}(C), \hat{A}\phi + B\psi \in H\}$$

by

$$\mathcal{A} = \begin{pmatrix} 0 & I & 0 \\ \hat{A} & B & \hat{K} \\ 0 & I & C \end{pmatrix}.$$

That is,

$$A\Phi = (\psi, \hat{A}\phi + B\psi + \hat{K}\eta, \psi + C\eta)$$

for $\Phi = (\phi, \psi, \eta) \in \mathcal{D}(\mathcal{A})$.

Thus the equation (1.20) can be formally written as

$$\dot{x}(t) = \mathcal{A}x(t) + \mathcal{F}(t),$$

where $x(t) = (E(t), \dot{E}(t), w(t))$ and $\mathcal{F} = (0, \mathcal{J}, 0)$.

1.7 Example 5: Structured Population Models

We consider the special case of "transport" models or Example 2 with $D = 0$ but with so-called renewal or recruitment boundary conditions. The "spatial" variable ξ is actually "size" in place of spatial location (see [BT]) and such models have been effectively used to model marine populations such as mosquitofish [BBKW, BF, BFPZ] and, more recently, shrimp [GRD-FP, BDEHAD, GRD-FP2]. Such models have also been the basis of labeled cell proliferation models in which ξ represents label intensity [BSTBRSM, BST2]. The early versions of these size structured population models were first proposed by Sinko and Streifer [SS] in 1967. Cell population versions were proposed by Bell and Anderson [BA] almost simultaneously. Other versions of these models called "age structured models," where age can be "physiological age" as well as chronological age, are discussed in [MetzD].

The *Sinko-Streifer model (SS)* for size-structured mosquitofish populations is given by

$$\frac{\partial v}{\partial t} + \frac{\partial}{\partial \xi}(gv) = -\mu v, \quad \xi_0 < \xi < \xi_1, \quad t > 0 \qquad (1.22)$$

$$v(0, \xi) = \Phi(\xi)$$

$$g(t, \xi_0)v(t, \xi_0) = \int_{\xi_0}^{\xi_1} K(t, s)v(t, s)ds$$

$$g(t, \xi_1) = 0.$$

Here $v(t, \xi)$ represents number density (given in numbers per unit length) or population density, where t represents time and ξ represents the length of the mosquitofish. The growth rate of individual mosquitofish is assumed given by $g(t, \xi)$, where

$$\frac{d\xi}{dt} = g(t, \xi) \qquad (1.23)$$

for each individual (all mosquitofish of a given size have the same growth rate).

In the SS model $\mu(t, \xi)$ represents the *mortality rate* of mosquitofish, and the function $\Phi(\xi)$ represents the initial size density of the population, while K represents the *fecundity kernel*. The boundary condition at $\xi = \xi_0$ is *recruitment*, or *birth rate*, while the boundary condition at $\xi = \xi_1 = \xi_{max}$ ensures the maximum size of the mosquitofish is ξ_1. The SS model *cannot* be used as formulated above to model the mosquitofish population because it *does not predict dispersion or bifurcation* of the population in time under biologically reasonable assumptions [BBKW, BF, BFPZ]. As we shall see, we will subsequently replace the growth rate g by a family $\mathcal{G}$ of growth rates and reconsider the model with a probability distribution P on this family. The population density is then given by summing "cohorts" of subpopulations where individuals belong to the same subpopulation if they have the same growth rate [BD, BDTR, GRD-FP, BDEHAD, GRD-FP2]. Thus, in the so-called Growth Rate Distribution (GRD) model, the population density $u(t, \xi; P)$, first discussed in [BBKW] and developed in [BF], is actually given by

$$u(t, \xi; P) = \int_{\mathcal{G}} v(t, \xi; g)dP(g), \qquad (1.24)$$

where $\mathcal{G}$ is a collection of admissible growth rates, P is a probability measure on $\mathcal{G}$, and $v(t, \xi; g)$ is the solution of the (SS) with growth rate g. This model assumes the population is made up of *collections of subpopulations* with individuals in the same subpopulation having the same size dependent growth rate. This example can also be formulated in terms of semigroups [BKa, BKW1, BKW2], but the details are somewhat more difficult than those for the first three examples. In some cases it is advantageous to use a *weak formulation* (to be developed later) instead of a semigroup formulation.

2 Semigroups and Infinitesimal Generators

Having introduced several examples of concrete distributed parameter systems in the previous chapter, we turn to a discussion of semigroups and their properties.

2.1 Basic Principles of Semigroups

Here we summarize results that are readily found (and proved) in other texts including [HP, Pa, Sh, T].

Definition 2.1 A *semigroup* is a one parameter set of operators $\{T(t) : T(t) \in \mathcal{L}(X)\}$, where X is a Banach or Hilbert space, such that $T(t)$ satisfies

1. $T(t+s) = T(t)T(s)$ (semigroup or Markov or translation property)

2. $T(0) = I$.   (identity property)

Classification of Semigroups by Continuity

- $T(t)$ is *uniformly continuous* if $\lim\limits_{t \to 0^+} |T(t) - I| = 0$.

  This is not of interest to us, because $T(t)$ is uniformly continuous if and only if $T(t) = e^{At}$ where A is a bounded linear operator.

- $T(t)$ is *strongly continuous*, denoted C_0, if for each $x \in X$, $t \to T(t)x$ is continuous on $[0, \delta]$ for some positive δ.

Note 1: All continuity statements are in terms of continuity from the right at zero. For fixed t

$$
\begin{aligned}
T(t+h) - T(t) &= T(t)[T(h) - T(0)] \\
&= T(t)[T(h) - I]
\end{aligned}
$$

and

$$
T(t) - T(t - \epsilon) = T(t - \epsilon)[T(\epsilon) - I]
$$

so that continuity from the right at zero is equivalent to continuity at any t for operators that are uniformly bounded on compact intervals.

Note 2: $T(t)$ uniformly continuous implies $T(t)$ strongly continuous ($|T(t)x - x| \leq |T(t) - I| \, |x|$), but not conversely.

2.2   Infinitesimal Generators

Definition 2.2 Assume $\{T(t) \in \mathcal{L}(X); 0 \leq t < \infty\}$ is a C_0 semigroup in X or on X, then the *infinitesimal generator* A of the semigroup $T(t)$ is defined by

$$\mathcal{D}(A) = \{x \in X : \lim_{t \to 0+} \frac{T(t)x - x}{t} \text{ exists in X}\}$$

and

$$Ax = \lim_{t \to 0+} \frac{T(t)x - x}{t}$$

with $A : \mathcal{D}(A) \subset X \to X$.

Theorem 2.1 *Let* $T(t)$ *be a* C_0 *semigroup in* X. *Then there exist constants* $\omega \geq 0$ *and* $M \geq 1$ *such that*

$$|T(t)| \leq Me^{\omega t} \qquad t \geq 0.$$

See [Pa]: Theorem 2.2.

Outline of proof: We want to show there exists $\eta > 0$ and $M > 0$ such that $|T(t)| \leq M$ on $[0, \eta]$. If not, then there exists $t_n \to 0^+$ such that $|T(t_n)| \geq n$, which implies $|T(t_n)x|$ is unbounded for some $x \in X$ (by the uniform boundedness principle). This contradicts the fact that for some positive δ, $t \to T(t)x$ is continuous on $[0, \delta]$ for each $x \in X$. We thus may define $\omega \equiv \frac{1}{\eta} \ln(M)$. Then $e^{\omega t} = M^{t/\eta}$ for any $t > 0$.

Let $t = \delta + k\eta$ for some integer k and $\delta \in [0, \eta]$. Then $|T(t)| = |T(\delta)T(\eta)^k| \leq MM^k \leq MM^{t/\eta} = Me^{\omega t}$.

Notation: Write $A \in G(M, \omega)$. For $M = 1$ and $\omega = 0$, $A \in G(1, 0)$ is the infinitesimal generator of the contraction semigroup: $|T(t)x - T(t)y| \leq |x - y|$.

Theorem 2.2 *Let* $T(t)$ *be a* C_0 *semigroup and let* A *be its infinitesimal generator. Then*

1. *For* $x \in X$,

$$\lim_{h \to 0} \frac{1}{h} \int_t^{t+h} T(s)x \, ds = T(t)x.$$

2. For $x \in X$,

$$\int_0^t T(s)x \, ds \in \mathcal{D}(A) \quad and \quad A\left(\int_0^t T(s)x \, ds\right) = T(t)x - x.$$

3. If $x \in \mathcal{D}(A)$, then $T(t)x \in \mathcal{D}(A)$ and

$$\frac{d}{dt}T(t)x = AT(t)x = T(t)Ax.$$

In other words, $\mathcal{D}(A)$ is invariant under $T(t)$, and on $\mathcal{D}(A)$ at least, $T(t)x_0$ is a solution of $\begin{cases} \dot{x}(t) &= Ax(t) \\ x(0) &= x_0. \end{cases}$

4. For $x \in \mathcal{D}(A)$,

$$T(t)x - T(s)x = \int_s^t T(\tau)Ax \, d\tau = \int_s^t AT(\tau)x \, d\tau.$$

See [Pa]: Theorem 2.4.

Corollary 2.1 $A \in G(M, \omega) \Rightarrow \mathcal{D}(A)$ *is dense in X and A is a* <u>*closed*</u> *linear operator.*

See [Pa]: Corollary 2.5.

Recall: By definition, a linear operator A is closed if and only if A has a closed graph in $X \times X$. That is, $Gr(A) = \{(x, y) \in X \times X \mid x \in \mathcal{D}(A), y = Ax\}$ is closed or for any $(x_n, y_n) \in Gr(A)$, if $x_n \to x$ and $y_n \to y$, then $x \in \mathcal{D}(A)$ and $y = Ax$.

Question: Does there exist a one-to-one relationship between a semigroup and its infinitesimal generator?

Theorem 2.3 *Let $T(t)$ and $S(t)$ be C_0 semigroups on X with infinitesimal generators A and B, respectively. If $A = B$, then $T(t) = S(t)$, i.e., the infinitesimal generator uniquely determines the semigroup on all of X.*

See [Pa]: Theorem 2.6.

3 Generators

3.1 Introduction to Generation Theorems

How do we tell when an operator A, derived from a PDE (recall Example 1: the heat equation), is actually a generator of a C_0 semigroup? This is important, because it leads to the idea of <u>well-posedness</u> and <u>continuous</u> dependence of solutions for an IBVPDE.

Well-posedness of a PDE is equivalent to saying that a unique solution exists in some sense and is continuous with respect to data. In other words,

$$\begin{cases} \dot{x}(t) & = & Ax(t) + F(t) \\ x(0) & = & x_0, \end{cases}$$

is satisfied in some sense and the corresponding semigroup generated solution $x(t) = T(t)x_0 + \int_0^t T(t-s)F(s)ds$, yields the map $(x_0, F) \to x(\cdot; x_0, F)$, that is then continuous in some sense (depending on the spaces used).

Probably the best-known generation theorem is the *Hille-Yosida* theorem. First, however, we review resolvents. In the study of C_0-semigroups, one frequently encounters the operators $\lambda I - A$, for $\lambda \in \mathbb{C}$ (also denoted by $\lambda - A$), and their inverse $R_\lambda(A) = (\lambda - A)^{-1}$. Here $\mathbb{C}$ is the field of complex scalars. For a linear operator A in X, we denote the *resolvent set* by $\rho(A) = \{\lambda \in \mathbb{C} | \lambda - A$ has range $\mathcal{R}(\lambda - A)$ dense in X and $\lambda - A$ has a continuous inverse on $\mathcal{R}(\lambda - A)\}$. We recall that any continuous densely defined linear operator in X can be extended continuously to all of X. For $\lambda \in \rho(A)$, we denote the resolvent operator in $\mathcal{L}(X)$ by $R_\lambda(A) = (\lambda - A)^{-1}$. The *spectrum* $\sigma(A)$ of a linear operator A is the complement in $\mathbb{C}$ of the resolvent set $\rho(A)$.

3.2 *Hille-Yosida* Theorems

We discuss first what are probably the most well-known and perhaps most basic generation theorems that can be found in numerous other texts including [HP, Pa, Sh, T].

Theorem 3.1 *(Hille-Yosida) For $M \geq 1, \omega \in \mathbb{R}$, we have $A \in G(M, \omega)$ if and only if*

1. *A is closed and densely defined (i.e., A closed and $\overline{D(A)} = X$).*

2. *For real $\lambda > \omega$, we have $\lambda \in \rho(A)$ and $R_\lambda(A)$ satisfies*

$$|R_\lambda(A)^n| \leq \frac{M}{(\lambda - \omega)^n}, \qquad n = 1, 2, \ldots.$$

Note: The equivalent version of *Hille-Yosida* in [Pa] is stated differently:

Theorem 3.2 *(Hille - Yosida) $A \in G(1,0)$ $\Longleftrightarrow$*

1. *A is closed and densely defined.*

2. $\mathbb{R}^+ \subset \rho(A)$ *and for every $\lambda > 0$, $|R_\lambda(A)| \leq \frac{1}{\lambda}$.*

See [Pa, Theorem 3.1]

Homework Exercises

- **Ex. 5** :

  a) Study the proof of Theorem 3.1 in [Pa] and read Section 1.3 carefully.

  b) Show that the version of *Hille-Yosida* in [Pa] is completely equivalent to the version stated previously. (Hint: This is not an exercise in proving *Hille-Yosida*. It is an exercise in comparing $A \in G(M, \omega)$ and $A \in G(1, 0)$ with regard to exponential rates and norms.)

3.3   Results from the *Hille-Yosida* Proof

Results from the Necessity Portion of the Proof

Out of the necessity portion of the proof of *Hille-Yosida* we find the representation:

$$R_\lambda(A)x = R(\lambda, A)x \equiv \int_0^\infty e^{-\lambda t} T(t)x \, dt \qquad \text{for } x \in X, \quad \lambda > \omega$$

This says that the Laplace transform of the semigroup is the resolvent operator.

We might ask the question: can we recover the semigroup $T(t)$ from the resolvent $R_\lambda(A)$ through some kind of inverse Laplace transform? The answer is yes! (see [Pa, p. 29-32],[DS, p. 646]) although we will not need or present these interesting results in this book.

Homework Exercises

- Ex. 6 : Show that the heat equation operator of Example 1 generates a C_0 semigroup in $X = L_2(0, l)$.

  (That is, show that $A\varphi = (D(\xi)\varphi'(\xi))'$ on $\mathcal{D}(A) = \{\varphi \in H^2(0, l) | \varphi(0) = 0, \varphi'(l) = 0\}$ is the infinitesimal generator of a C_0 semigroup in X.)

  Note: If you want to, you can take $D(\xi) = D$ constant for this exercise.

Results from the Sufficiency Portion of the Proof

In the sufficiency proof of *Hille-Yosida*, we encounter the Yosida approximation:
For $\lambda > \omega$,

$$A_\lambda \equiv \lambda A R_\lambda(A) = \lambda^2 R_\lambda(A) - \lambda I.$$

From this definition, we can see that A_λ is bounded and defined on all of X. To see the above relationship, note that

$$\lambda A R_\lambda(A) = \lambda[(A - \lambda I)R_\lambda(A) + \lambda R_\lambda(A)].$$

We also find that for A satisfying the hypothesis of *Hille-Yosida*, we have

$$\lim_{\lambda \to \infty} A_\lambda x = Ax \qquad \text{for all } x \in \mathcal{D}(A).$$

This says that for large λ, A_λ acts like A. Moreover, under this hypothesis, A_λ is the infinitesimal generator of the uniformly continuous semigroup of contraction operators : e^{tA_λ}. Then we can argue that $T(t)x \equiv \lim_{\lambda \to \infty} e^{tA_\lambda}x$, for all $x \in X$, is the desired semigroup that A generates. (The proof is constructive.) See Corollary 3.5 in [Pa].

3.4 Corollaries to *Hille-Yosida*

Corollary 3.1 *Let A be the infinitesimal generator of a C_0 semigroup in X, and A_λ be the Yosida approximation, then*

$$T(t)x = \lim_{\lambda \to \infty} e^{tA_\lambda}x \qquad \text{for all } x \in X.$$

Corollary 3.2 *If A is the infinitesimal generator of a C_0 semigroup in X, then we actually have $\{\lambda \in \mathbb{C}| Re\lambda > \omega\} \subset \rho(A)$ and for such λ*

$$|R_\lambda(A)^n| \le \frac{M}{(Re\lambda - \omega)^n}.$$

Corollary 3.3 *$A \in G(M, \omega)$ implies $\mathcal{D}(A)$ is dense in X, and moreover, we find $\bigcap\limits_{n=1}^{\infty} \mathcal{D}(A^n)$ is also dense in X.*

3.5   *Lumer-Phillips* and Dissipative Operators

While the Hille-Yosida results above are classical and well known, in numerous applications there are other generation theorems and subsequent corollaries that are more easily used. These require the underlying system be described by a *dissipative* operator.

Definition 3.1 Let X be a Hilbert space. Then a linear operator $A : \mathcal{D}(A) \subset X \to X$ is *dissipative* if

$$Re \langle Ax, x\rangle \le 0 \quad \text{for all } x \in \mathcal{D}(A).$$

Theorem 3.3 *A linear operator A is dissipative if and only if $|(\lambda I - A)x| \ge \lambda|x|$ for all $x \in \mathcal{D}(A)$ and $\lambda > 0$.*

This result yields that for dissipative operators we have $\lambda I - A$ is continuously invertible on $\mathcal{R}(\lambda - A)$ when $\lambda > 0$. Thus, $\lambda > 0$ implies $\lambda \in \rho(A)$ whenever $\mathcal{R}(\lambda - A)$ is dense in X.

Theorem 3.4 *(Lumer-Phillips) Suppose A is a linear operator in a Hilbert space X.*

1. *If A is densely defined, $A - \omega I$ is dissipative for some real ω, and $\mathcal{R}(\lambda_0 - A) = X$ for some λ_0 with $Re \lambda_0 > \omega$, then $A \in G(1, \omega)$.*

2. *If $A \in G(1, \omega)$, then A is densely defined, $A - \omega I$ is dissipative and $\mathcal{R}(\lambda_0 - A) = X$ for all λ_0 with $Re \lambda_0 > \omega$.*

Proof of Theorem 3.3

A is dissipative means Re $\langle Ax, x \rangle \leq 0$ for all $x \in \mathcal{D}(A)$. This implies Re$\langle -Ax, x \rangle \geq 0$. So we have:

$$
\begin{aligned}
|(\lambda - A)x||x| &\geq |\langle (\lambda - A)x, x \rangle| \\
&\geq \text{Re}\langle (\lambda - A)x, x \rangle \\
&\geq \lambda \langle x, x \rangle \\
&= \lambda |x|^2.
\end{aligned}
$$

Conversely, suppose $|(\lambda I - A)x| \geq \lambda |x|$ for all $x \in \mathcal{D}(A)$ and $\lambda > 0$. Let $x \in \mathcal{D}(A)$. Define $y_\lambda = (\lambda - A)x$ and $z_\lambda = \frac{y_\lambda}{|y_\lambda|}$. Then we have $|z_\lambda| = 1$ and

$$
\begin{aligned}
\lambda |x| &\leq |(\lambda - A)x| \\
&= \langle (\lambda - A)x, z_\lambda \rangle \\
&= \lambda \text{Re}\langle x, z_\lambda \rangle - \text{Re}\langle Ax, z_\lambda \rangle \\
&\leq \lambda |x||z_\lambda| - \text{Re}\langle Ax, z_\lambda \rangle \\
&= \lambda |x| - \text{Re}\langle Ax, z_\lambda \rangle.
\end{aligned}
\tag{3.1}
$$

This implies

$$
\text{Re}\langle Ax, z_\lambda \rangle \leq 0.
\tag{3.2}
$$

We always have the relationship

$$
-\text{Re}\langle Ax, z_\lambda \rangle \leq |Ax|.
\tag{3.3}
$$

Combining equations (3.1) and (3.3), we obtain

$$
\begin{aligned}
\lambda |x| &\leq \lambda \text{Re}\langle x, z_\lambda \rangle - \text{Re}\langle Ax, z_\lambda \rangle \\
&\leq \lambda \text{Re}\langle x, z_\lambda \rangle + |Ax|.
\end{aligned}
\tag{3.4}
$$

Rearranging equation (3.4), we have

$$
\lambda \text{Re}\langle x, z_\lambda \rangle \geq \lambda |x| - |Ax|
$$

or

$$
\text{Re}\langle x, z_\lambda \rangle \geq |x| - \frac{1}{\lambda}|Ax|.
\tag{3.5}
$$

Observe that we have $|z_\lambda| = 1$ for $\lambda > 0$. As we shall see below in Chapter 4, this implies that for some subsequence $\{z_{\lambda_k}\}$ and some $\{\tilde{z}\}$ with $|\tilde{z}| \leq 1$, we have $\langle z_{\lambda_k}, x \rangle \to \langle \tilde{z}, x \rangle$ as $\lambda_k \to \infty$ for all $x \in X$. This is called weak convergence and is denoted by $z_{\lambda_k} \rightharpoonup \tilde{z}$. (Note: The norm is weakly lower semi-continuous; therefore, $z_{\lambda_k} \rightharpoonup \tilde{z}$ implies $|\tilde{z}| = |\text{weak}\lim_k z_{\lambda_k}| \leq \underline{\lim}_k |z_{\lambda_k}| = 1$. This is discussed in detail in Section 4.2.)

Taking the limit of equation (3.2), we obtain

$$\text{Re}\langle Ax, \tilde{z} \rangle \leq 0. \tag{3.6}$$

Similarly, taking the limit of equation (3.5), we have

$$\text{Re}\langle x, \tilde{z} \rangle \geq |x|. \tag{3.7}$$

Thus because $|\tilde{z}| \leq 1$ we can make the arguments

$$
\begin{aligned}
|x| &\leq \text{Re}\langle x, \tilde{z} \rangle \\
&\leq |\langle x, \tilde{z} \rangle| \\
&\leq |x||\tilde{z}| \\
&\leq |x|.
\end{aligned}
$$

Therefore, we actually have the equality

$$\text{Re}\langle x, \tilde{z} \rangle = |\langle x, \tilde{z} \rangle| = \langle x, \tilde{z} \rangle = |x|. \tag{3.8}$$

Define $\tilde{x} = \tilde{z}|x|$; then $\langle x, \tilde{x} \rangle = |x|^2$ by (3.8) as $\langle x, \tilde{z} \rangle = |x|$ implies $\langle x, \tilde{z}|x| \rangle = |x|\langle x, \tilde{z} \rangle = |x|^2$. From equation (3.6),

$$\text{Re}\langle Ax, \tilde{z} \rangle \leq 0$$

which implies

$$\text{Re}\langle Ax, \tilde{z}|x| \rangle \leq 0$$

or

$$\text{Re}\langle Ax, \tilde{x} \rangle \leq 0. \tag{3.9}$$

We also have $|\tilde{x}| \leq |x|$ because $|\tilde{x}| = |\tilde{z}||x| \leq |x|$.

We claim that $\tilde{x} = x$. If this is true, then (3.9) would imply $\text{Re}\langle Ax, x \rangle \leq 0$ for all $x \in \mathcal{D}(A)$ and we are finished. We need the following lemma to prove this.

Lemma 3.1 *(Duality Set Lemma) Let $x \in X$ be arbitrary in a Hilbert space X and $\tilde{x} \in X$ such that $\langle \tilde{x}, x \rangle = |x|^2$ and $|\tilde{x}| \leq |x|$; then $\tilde{x} = x$.*

Proof

$$
\begin{aligned}
0 &\leq \langle x - \tilde{x}, x - \tilde{x} \rangle \\
&= |x|^2 - \langle \tilde{x}, x \rangle - \langle \tilde{x}, x \rangle + |\tilde{x}|^2 \\
&\leq |x|^2 - |x|^2 - |x|^2 + |x|^2 \\
&= 0
\end{aligned}
$$

Therefore, $|x - \tilde{x}| = 0$ and $x = \tilde{x}$; i.e., in a Hilbert space, $F(x) = \{x\}$, where for $x \in X$, the duality set $F(x) \subseteq X^*$ is given by $F(x) \equiv \{x^* \in X | \langle x, x^* \rangle = |x|^2 = |x^*|^2\}$.

As we shall see below, for a general Banach space X, the duality set is a subset of the dual space X^*. Because Hilbert spaces are self-dual, in a Hilbert space X the concept reduces to one involving subsets of the Hilbert space X itself.

Definition 3.2 Let X be a Hilbert space with $A : \mathcal{D}(A) \subset X \to X$ with $\mathcal{D}(A)$ dense in X. Then the *adjoint of A*, denoted A^*, is defined by

$$
\begin{aligned}
\mathcal{D}(A^*) = \{y \in X | \text{ there exists } w \in X \text{ satisfying } \langle Ax, y \rangle \\
= \langle x, w \rangle \text{ for all } x \in \mathcal{D}(A)\}
\end{aligned}
$$

and

$$
A^* y = w.
$$

Corollary 3.4 *(to Lumer-Phillips) Let X be a Hilbert space and $A : \mathcal{D}(A) \subset X \to X$ with $\mathcal{D}(A)$ dense and A closed in X. If both A and A^* are dissipative, then A is an infinitesimal generator of a C_0 semigroup of contractions on X.*

Proof By Lumer-Phillips, it suffices to argue that $\mathcal{R}(\lambda - A) = X$ for some $\lambda > 0$.

Take $\lambda = 1$:

Recall that in a Hilbert space we have

$$
\langle Ax, y \rangle = \langle x, A^* y \rangle \text{ for } x \in \mathcal{D}(A), y \in \mathcal{D}(A^*).
$$

But by assumption A is dissipative and closed, which means

$$
\{(x, Ax) | x \in \mathcal{D}(A)\} \text{ is closed in } X \times X.
$$

Therefore, $\mathcal{R}(I - A)$ is a closed subspace of X.

Claim: $\mathcal{R}(I - A) = X$.

Suppose $\mathcal{R}(I - A) \neq X$. Then by the Hahn-Banach theorem in a Hilbert space, there exists $\tilde{x} \in X, \tilde{x} \neq 0$, such that

$$\langle \tilde{x}, (I - A)x \rangle = 0.$$

In other words

$$\langle \tilde{x}, x \rangle = \langle \tilde{x}, Ax \rangle \quad \text{for all } x \in \mathcal{D}(A).$$

Therefore

$$\tilde{x} \in \mathcal{D}(A^*) \text{ and } \langle A^*\tilde{x}, x \rangle = \langle \tilde{x}, x \rangle,$$

or

$$\langle \tilde{x} - A^*\tilde{x}, x \rangle = 0 \quad \text{for all } x \in \mathcal{D}(A).$$

However, $\mathcal{D}(A)$ is dense, which implies

$$\tilde{x} - A^*\tilde{x} = 0 \text{ or } \tilde{x} = A^*\tilde{x}. \tag{3.10}$$

By assumption, A^* is also dissipative. Therefore

$$|(I - A^*)\tilde{x}| \geq |\tilde{x}| \quad \text{for all } \tilde{x} \in \mathcal{D}(A^*). \tag{3.11}$$

Combining (3.10) and (3.11) we have

$$\begin{array}{rcl} |\tilde{x}| & \leq & |\tilde{x} - A^*\tilde{x}| \\ |\tilde{x}| & \leq & |\tilde{x} - \tilde{x}| \\ |\tilde{x}| & \leq & 0. \end{array}$$

This is a contradiction since $\tilde{x} \neq 0$. Therefore, $\mathcal{R}(I - A) = X$.

Theorem 3.5 *In a Hilbert space X, if A is dissipative and $\mathcal{R}(\lambda - A) = X$ for some $\lambda > 0$, then A is densely defined.*

See [Pa, Theorem 4.6].

Note: This also holds in a reflexive Banach space (more on this later).

Important Practical Results:

1. In a Hilbert space, we only need A to be dissipative and $\mathcal{R}(\lambda I - A) = X$ for some $\lambda > 0$ in order to use the Lumer-Phillips theorem.

2. In a real Hilbert space (such as real $L_2(0, l)$), we can get by with $\langle Ax, x \rangle \leq 0$.

Lemma 3.2 *(A Practical Lemma) Let X be a complex Hilbert space and suppose that for $u \in \mathcal{D}(A)$, Au is real whenever u is real (for example, an operator A coming from a PDE defined by real coefficients), then A is dissipative, i.e.,*

$$Re\langle Ax, x \rangle \leq 0 \quad \textit{for all } x \in \mathcal{D}(A)$$

if and only if

$$\langle Au, u \rangle \leq 0 \quad \textit{for all real valued } u \in \mathcal{D}(A).$$

Proof:
   We have for $x = u + iv$, with u, v, real:

$$
\begin{aligned}
\langle Ax, x \rangle &= \langle A(u + iv), u + iv \rangle \\
&= \langle Au, u \rangle + \langle Av, v \rangle + i[\langle Av, u \rangle - \langle Au, v \rangle].
\end{aligned}
$$

We want to argue that $\langle Av, u \rangle - \langle Au, v \rangle$ is real; however, for any complex Hilbert space we have the *polarization identity* [GP, p. 168]

$$\langle u, v \rangle = \frac{1}{4}\{|u + v|^2 + i|u + iv|^2 - i|u - iv|^2 - |u - v|^2\}.$$

But if u and v are real-valued we have $|u + iv| = |u - iv|$ and hence

$$\langle u, v \rangle = \frac{1}{4}\{|u + v|^2 - |u - v|^2\}$$

(note that this is the polarization identity in a real Hilbert space) which is real valued. Since u and v are real, then by assumption Au and Av are real. Therefore, $\langle Av, u \rangle - \langle Au, v \rangle$ is real-valued and hence

$$Re \langle Ax, x \rangle = \langle Au, u \rangle + \langle Av, v \rangle.$$

Thus to have Re $\langle Ax, x \rangle \leq 0$, it suffices to argue

$$\langle Au, u \rangle \leq 0 \text{ for all real valued } u \text{ in } \mathcal{D}(A).$$

Remark: Any Banach space is a Hilbert space if and only if the parallelogram law

$$|x + y|^2 + |x - y|^2 = 2|x|^2 + 2|y|^2$$

holds. If this holds in a Banach space, the polarization identity can be used to define a norm compatible inner product.

To illustrate the use of the Lumer-Phillips theorem, we return to some of the previous examples.

3.6   Examples Using Lumer-Phillips Theorem

3.6.1   Return to Example 1: The Heat Equation

We return to the heat diffusion example where we found the PDE

$$\frac{\partial y}{\partial t}(t, \xi) = \frac{\partial}{\partial \xi}\left(D(\xi)\frac{\partial y}{\partial \xi}\right) + f(t, \xi)$$

$$y(t, 0) = 0, \quad D(l)\frac{\partial y}{\partial \xi}(t, l) = 0$$

gave rise to an operator A in $X = L_2(0, l)$ given by

$$\mathcal{D}(A) = \{\varphi \in H^2(0, l) | (D\varphi')' \in L_2(0, l), \varphi(0) = 0, \varphi'(l) = 0\},$$

and

$$A\varphi = (D\varphi')'.$$

Remark: Notice that if D is smooth, we have

$$\mathcal{D}(A) = \{\varphi \in H^2(0, l) | \varphi(0) = 0, \varphi'(l) = 0\}.$$

In view of the remarks following the Lumer-Phillips theorem, to show that A generates a C_0 semigroup in X, it suffices to show that A is dissipative and that $\mathcal{R}(\lambda - A) = X$; i.e. $(\lambda - A)\mathcal{D}(A) = X$ for some $\lambda > 0$.

We have for $\varphi \in D(A)$, φ real-valued:

$$
\begin{aligned}
\langle A\varphi, \varphi \rangle &= \langle (D\varphi')', \varphi \rangle \\
&= \int_0^l (D\varphi')'\varphi \\
&= -\langle D\varphi', \varphi' \rangle + D\varphi'\varphi|_0^l \\
&= -\int_0^l D(\varphi')^2 \\
&\leq 0
\end{aligned}
$$

for $D \geq 0$ (a natural assumption since D is the coefficient of thermal diffusion).

To argue that $R(\lambda - A) = L_2(0, l)$ for some $\lambda > 0$, we note that this is equivalent to arguing that for some $\lambda > 0$, the boundary value problem

$$
\lambda \varphi - (D\varphi')' = \psi, \quad 0 < x < l,
$$
$$
\varphi(0) = 0, \quad \varphi'(l) = 0,
$$

has a solution $\varphi \in H^2(0, l)$ for any $\psi \in L_2(0, l)$. But one can answer this latter question in the affirmative (assuming $D > 0$ is sufficiently regular, e.g., $D \in W^{1,\infty}$) using classical Sturm-Liouville theory. (See [BR, CH, D, S].) Thus, we readily see that A is a densely-defined generator of a C_0 solution semigroup for the system above.

Homework Exercises

- Ex. 7 : Consider the general transport example, Example 2, and show that it generates a C_0 semigroup.

3.6.2 Return to Example 2: The General Transport Equation

We next reconsider the population model given by

$$
\frac{\partial y}{\partial t}(t, \xi) + \frac{\partial}{\partial \xi}(\nu(\xi)y(t, \xi)) = \frac{\partial}{\partial \xi}\left(D(\xi)\frac{\partial y}{\partial \xi}(t, \xi)\right) - \mu(\xi)y(t, \xi)
$$

$$
y(t, 0) = 0, \quad \left[D(\xi)\frac{\partial y}{\partial \xi}(t, \xi) - \nu(\xi)y(t, \xi)\right]_{\xi=l} = 0.
$$

The corresponding operator equation in $X = L_2(0, l)$ is in terms of the operator A given by

$$
D(A) =
$$
$$
\{\varphi \in H^2(0, l) \,|\, (D\varphi')' - (\nu\varphi)' \in L_2(0, l), \varphi(0) = 0, \left[D\varphi' - \nu\varphi\right](l) = 0\}
$$

$$A\varphi = (D\varphi')' - (\nu\varphi)' - \mu\varphi.$$

Again we show that A is dissipative in $L_2(0, l)$ (actually we shall argue that a translation of A is dissipative). For real valued $\varphi \in \mathcal{D}(A)$, we have

$$
\begin{aligned}
\langle A\varphi, \varphi \rangle &= \langle (D\varphi' - \nu\varphi)', \varphi \rangle - \langle \mu\varphi, \varphi \rangle, \\
&= -\langle (D\varphi' - \nu\varphi), \varphi' \rangle + (D\varphi' - \nu\varphi)\varphi|_0^l - \langle \mu\varphi, \varphi \rangle, \\
&= -\langle D\varphi', \varphi' \rangle - \langle \mu\varphi, \varphi \rangle + \langle \nu\varphi, \varphi' \rangle.
\end{aligned}
$$

Hence, using $ab = \frac{1}{\sqrt{2\epsilon}} a \sqrt{2\epsilon} b \le \frac{1}{4\epsilon} a^2 + \epsilon b^2$, we find

$$
\begin{aligned}
\langle A\varphi, \varphi \rangle &\le -\langle D\varphi', \varphi' \rangle - \langle \mu\varphi, \varphi \rangle + \frac{1}{4\epsilon}|\nu\varphi|^2 + \epsilon|\varphi'|^2 \\
&= \langle (-D + \epsilon)\varphi', \varphi' \rangle + \langle (-\mu + \frac{\nu^2}{4\epsilon})\varphi, \varphi \rangle \\
&\le \frac{k}{4\epsilon}|\varphi|^2
\end{aligned}
$$

for ϵ sufficiently small so that $D(\xi) \ge \epsilon$ and k chosen so that $-4\epsilon\mu + \nu^2(\xi) \le k$. Thus, we have that $A - (\frac{k}{4\epsilon})I$ is dissipative in $L_2(0, l)$.

The range statement $\mathcal{R}(\lambda + \frac{k}{4\epsilon} - A) = X$ again reduces to a question for Sturm-Liouville problems. For any $\psi \in L_2(0, l)$, we seek an $L_2(0, l)$ solution of

$$(\lambda + \frac{k}{4\epsilon} + \mu)\varphi + (\nu\varphi)' - (D\varphi')' = \psi, \quad 0 < \xi < l,$$

$$\varphi(0) = 0, \quad [D\varphi' - \nu\varphi](l) = 0.$$

For sufficiently large λ, the Sturm-Liouville theory guarantees existence of solutions. Thus $A - \frac{k}{4\epsilon}$ is the generator of a C_0-semigroup $S(t)$ so that $T(t) = e^{\frac{k}{4\epsilon}t}S(t)$ is the solution semigroup generated by A (this follows from the translation properties of differential equation solutions and Theorem 2.2.

3.6.3   Return to Example 3: Delay Systems

We return to the delay system equation of Example 3 which is a special case of more general vector delay systems. We consider the Cauchy problem

$$\dot{z}(t) = L(z_t), \quad (z(0), z_0) = (\eta_0, \phi_0), \tag{3.12}$$

where $z_t(\xi) = z(t + \xi)$, $\xi \in [-r, 0]$ and

$$L(z_t) = \sum_{i=0}^{m} A_i z(t - \tau_i) + \int_{-r}^{0} K(\xi) z(t + \xi) d\xi,$$

where $0 = \tau_0 < \tau_1 < \cdots < \tau_m = r$, and A_i and $K(\xi)$ are $n \times n$ matrices. For general functions $\phi \in C(-r, 0; \mathbb{R}^n)$, the mapping $L(\cdot) : C(-r, 0; \mathbb{R}^n) \to \mathbb{R}^n$ has the form

$$L(\phi) = \sum_{i=0}^{m} A_i \phi(-\tau_i) + \int_{-r}^{0} K(\xi) \phi(\xi) d\xi. \tag{3.13}$$

We note that L is only defined for functions ϕ for which pointwise evaluation is meaningful, for example, for $\phi \in C(-r, 0; \mathbb{R}^n) \subset L_2(-r, 0; \mathbb{R}^n)$, even though we will consider solutions in the Hilbert space $\mathbb{R}^n \times L_2(-r, 0; \mathbb{R}^n)$.

Let $x(t) = (z(t), z_t) \in X$, $X \equiv \mathbb{R}^n \times L_2(-r, 0; \mathbb{R}^n)$, where the Hilbert space X has the inner product

$$< (\eta, \phi), (\zeta, \psi) >_X = < \eta, \zeta >_{\mathbb{R}^n} + \int_{-r}^{0} < \phi(\xi), \psi(\xi) >_{\mathbb{R}^n} d\xi. \tag{3.14}$$

Define $\mathcal{A} : \mathcal{D}(\mathcal{A}) \in X \to X$, where $\mathcal{D}(\mathcal{A}) = \{\hat{\phi} = (\phi(0), \phi) \in X | \phi \in H^1(-r, 0; \mathbb{R}^n)\}$, which is dense in X. Then $\mathcal{A}\hat{\phi} = \mathcal{A}(\phi(0), \phi) = (L(\phi), \phi')$ for $\hat{\phi} = (\phi(0), \phi) \in \mathcal{D}(\mathcal{A})$. We use Lumer-Phillips to show that $\mathcal{A}$ is an infinitesimal generator of the C_0 semigroup $S(t)$ where $S(t)(\eta, \phi) = (z(t), z_t)$ for solutions of (3.12). We do this in a space X_g that is topologically equivalent to X, so that we have the semigroup on X as well as on X_g.

First, we show that $\mathcal{A}$ is dissipative in a space topologically equivalent to X. Renorm X by the weighting function g defined on $[-r, 0]$, where $g(\xi) = j$ for $\xi \in [-\tau_{m-j+1}, -\tau_{m-j})$, $j = 1, 2, \ldots, m$. Define the Hilbert space $X_g \equiv \mathbb{R}^n \times L_2(-r, 0; g; \mathbb{R}^n)$ to be the elements of X with this new inner product

$$< (\eta, \phi), (\zeta, \psi) >_{X_g} = < \eta, \zeta >_{\mathbb{R}^n} + \int_{-r}^{0} < \phi(\xi), \psi(\xi) >_{\mathbb{R}^n} g(\xi) d\xi.$$
$$\tag{3.15}$$

This gives rise to an equivalent topology to that of X as long as $g(\xi) > 0$ for all $\xi \in [-r, 0]$ (see [BBu2, p. 186], [BKap1, Webb] for more details). Then $\mathcal{A}$ is dissipative if

$$< \mathcal{A}x, x >_{X_g} \le \omega |x|^2_{X_g} \qquad (3.16)$$

for all $x \in \mathcal{D}(\mathcal{A})$ where $x = (\phi(0), \phi)$. To show this, we argue

$$
\begin{aligned}
< \mathcal{A}x, x > \;&=\; < (L(\phi), \phi'), (\phi(0), \phi) > \\
&=\; < L(\phi), \phi(0) >_{\mathbb{R}^n} + < \phi', \phi >_{L_2(-r,0;g;\mathbb{R}^n)} \\
&=\; < \sum_{i=0}^{m} A_i \phi(-\tau_i) + \int_{-r}^{0} K(\xi)\phi(\xi)d\xi, \phi(0) >_{\mathbb{R}^n} (3.17) \\
&\quad + \int_{-r}^{0} < \phi'(\xi), \phi(\xi) >_{\mathbb{R}^n} g(\xi)d\xi \\
&=\; < \sum_{i=0}^{m} A_i \phi(-\tau_i) + \int_{-r}^{0} K(\xi)\phi(\xi)d\xi, \phi(0) >_{\mathbb{R}^n} (3.18) \\
&\quad + \sum_{j=1}^{m} \int_{-\tau_{m-j+1}}^{-\tau_{m-j}} < \phi'(\xi), \phi(\xi) >_{\mathbb{R}^n} g(\xi)d\xi. \qquad (3.19)
\end{aligned}
$$

Consider the last term and denote $\phi_{m-j} = \phi(-\tau_{m-j})$

$$
\sum_{j=1}^{m} \int_{-\tau_{m-j+1}}^{-\tau_{m-j}} < \phi'(\xi), \phi(\xi) >_{\mathbb{R}^n} g(\xi)d\xi
$$

$$
= \sum_{j=1}^{m} \frac{1}{2} j |\phi(\xi)|^2 \big|_{\xi=-\tau_{m-j+1}}^{\xi=-\tau_{m-j}}
$$

$$
= \sum_{j=1}^{m} \left(\frac{j}{2} |\phi_{m-j}|^2 - \frac{j}{2} |\phi_{m-j+1}|^2 \right)
$$

$$
= \frac{1}{2} \sum_{j=1}^{m} j |\phi_{m-j}|^2 - \frac{1}{2} \sum_{k=0}^{m-1} (k+1) |\phi_{m-k}|^2 \quad (\text{ for } k = j - 1)
$$

$$
= \frac{1}{2} m |\phi_0|^2 + \frac{1}{2} \sum_{j=1}^{m-1} j |\phi_{m-j}|^2
$$

$$
- \frac{1}{2} \sum_{k=1}^{m-1} (k+1) |\phi_{m-k}|^2 - \frac{1}{2} |\phi_m|^2
$$

$$= \frac{1}{2}m|\phi_0|^2 - \frac{1}{2}\sum_{j=1}^{m-1}|\phi_{m-j}|^2 - \frac{1}{2}|\phi_m|^2$$

$$= \frac{1}{2}m|\phi_0|^2 - \frac{1}{2}\sum_{j=0}^{m-1}|\phi_{m-j}|^2$$

$$= \frac{m+1}{2}|\phi_0|^2 - \frac{1}{2}\sum_{j=0}^{m}|\phi_{m-j}|^2$$

Returning to (3.19), we have for $x = (\phi(0), \phi)$,

$$< \mathcal{A}x, x >_{X_g} \leq \sum_{i=0}^{m} |A_i||\phi_i||\phi_0| + < \int_{-r}^{0} K(\xi)\phi(\xi)d\xi, \phi(0) >_{\mathbb{R}^n}$$

$$+ \frac{m+1}{2}|\phi_0|^2 - \frac{1}{2}\sum_{j=0}^{m}|\phi_{m-j}|^2$$

$$\leq \sum_{i=0}^{m} \left(\frac{1}{2}|A_i|^2|\phi_0|^2 + \frac{1}{2}|\phi_i|^2\right) + |\int_{-r}^{0} K(\xi)\phi(\xi)d\xi||\phi(0)|$$

$$+ \frac{m+1}{2}|\phi_0|^2 - \frac{1}{2}\sum_{j=0}^{m}|\phi_{m-j}|^2$$

$$= \frac{1}{2}\sum_{i=0}^{m} |A_i|^2|\phi_0|^2 + \frac{1}{2}\sum_{i=0}^{m}|\phi_i|^2 + |\int_{-r}^{0} K(\xi)\phi(\xi)d\xi||\phi(0)|$$

$$+ \frac{m+1}{2}|\phi_0|^2 - \frac{1}{2}\sum_{j=0}^{m}|\phi_{m-j}|^2$$

$$\leq \left(\frac{1}{2}\sum_{i=0}^{m} |A_i|^2 + \frac{m+1}{2}\right)|\phi_0|^2 + \frac{1}{2}|K|_{L_2}^2|\phi|_{L_2}^2 + \frac{1}{2}|\phi(0)|^2$$

$$\leq \left(\frac{1}{2}\sum_{i=0}^{m} |A_i|^2 + \frac{m+1}{2} + \frac{1}{2} + \frac{1}{2}|K|_{L_2}^2\right)|x|_X^2$$

$$\leq \omega|x|_X^2$$

$$\leq \omega|x|_{X_g}^2,$$

using the definition of the inner product and the parallelogram identity; therefore choosing

$$\omega = \frac{1}{2} \sum_{i=0}^{m} |A_i|^2 + \frac{m}{2} + 1 + \frac{1}{2} |K|_{L_2}^2,$$

we have that $\mathcal{A}$ is dissipative in X_g.

To use Lumer-Phillips, it only remains to be shown that for some $\lambda > 0$, given any $\hat{\psi} = (\psi(0), \psi) \in X$, $(\lambda I - A)\hat{\phi} = \hat{\psi}$, can be solved for some $\hat{\phi} = (\phi(0), \phi) \in \mathcal{D}(\mathcal{A})$. This is equivalent to solving $\lambda\phi(0) - L(\phi) = \psi(0)$ and $\lambda\phi - \phi' = \psi$ on $[-r, 0]$. But this follows immediately from [BBu2, Lemma 3.3, p. 178].

3.6.4   Return to Example 4: Maxwell's Equations

We return to the Maxwell's equation in Example 4; specifically we consider the special case of (1.20). For our abstract operator formulation, we considered (1.20) in the state space $X = V \times H \times W = H_R^1(\Omega) \times L_2(\Omega) \times L_2(-r, 0; \tilde{g}; H)$ with states $(\phi, \psi, \eta) = (E(t), \dot{E}(t), w(t)) = (E(t), \dot{E}(t), E(t) - E(t + \cdot))$. To define an infinitesimal generator, we began by defining a fundamental set of component operators. First we formally defined $\hat{A} \in \mathcal{L}(V, V^*)$ by (see (1.21))

$$\hat{A}\phi \equiv c^2\phi'' - (\beta + g_{11})\phi + c^2\phi'(0)\delta_0 \tag{3.20}$$

where δ_0 is the Dirac operator $\delta_0\psi = \psi(0)$. More precisely, this means

$$\langle -\hat{A}\phi, \psi \rangle_{V^*, V} \equiv (-\hat{A}\phi)(\psi) = \langle c^2\phi', \psi' \rangle_H + \langle (\beta + g_{11})\phi, \psi \rangle_H \tag{3.21}$$

so that it is readily seen that $\hat{\sigma}_1 : V \times V \mapsto \mathbb{C}$ defined by

$$\hat{\sigma}_1(\phi, \psi) = \langle -\hat{A}\phi, \psi \rangle_{V^*, V} \tag{3.22}$$

is symmetric, V continuous (there is a γ such that $|\sigma(\phi, \psi)| \leq \gamma|\phi||\psi|$; this will be discussed in more detail later in Section 5), and satisfies $\hat{\sigma}_1(\phi, \phi) \geq c_1|\phi|_V^2 - \lambda_0|\phi|_H^2$ for constants λ_0 and $c_1 > 0$ (as we shall see below this is the defining property of V-coercive sesquilinear forms).

We also defined operators $B \in \mathcal{L}(V, V^*)$ and $\hat{K} \in \mathcal{L}(W, H)$ by

$$B\phi = -\gamma\phi - c\phi(0)\delta_0, \tag{3.23}$$

so that

$$\langle -B\phi, \psi \rangle_{V^*, V} = \langle \gamma\phi, \psi \rangle_H + c\phi(0)\psi(0) \tag{3.24}$$

and, for $\eta \in W = L^2(-r,0;\tilde{g};H)$,

$$\hat{K}\eta(z) = \int_{-r}^{0} \tilde{g}(\xi)\eta(\xi,z)d\xi \quad \text{for} \quad z \in \Omega. \tag{3.25}$$

We introduced another operator

$$C : \mathcal{D}(C) \subset W \mapsto W$$

defined on $\mathcal{D}(C) = \{\eta \in H^1(-r,0;H)|\eta(0) = 0\}$ by

$$C\eta(\theta) = \frac{\partial}{\partial\theta}\eta(\theta).$$

We then obtained the first-order system for $x(t)$ given by

$$\dot{x}(t) = \mathcal{A}x(t) + \mathcal{F}(t), \tag{3.26}$$

where $\mathcal{A}$ is given by

$$\mathcal{A} = \begin{pmatrix} 0 & I & 0 \\ \hat{A} & B & \hat{K} \\ 0 & I & C \end{pmatrix}$$

and

$$\mathcal{D}(\mathcal{A}) = \left\{ (\phi,\psi,\eta) \in X | \psi \in V, \eta \in \mathcal{D}(C), \hat{A}\phi + B\psi \in H \right\},$$

that is, $\mathcal{A}\Phi = (\psi, \hat{A}\phi + B\psi + \hat{K}\eta, \psi + C\eta)$ for $\Phi = (\phi,\psi,\eta)$. The forcing function $\mathcal{F}$ in (3.26) is given by $\mathcal{F} = (0,\mathcal{J},0)$. To argue that $\mathcal{A}$ is the infinitesimal generator of a C_0-semigroup, we actually consider the system (3.26) in an equivalent space $X_1 = V_1 \times H \times W$, where V_1 is the space V with equivalent inner product $\langle\phi_1,\phi_2\rangle_{V_1} = \hat{\sigma}_1(\phi_1,\phi_2)$, where $\hat{\sigma}_1$ is the sesquilinear form given in (3.22). Observe that $\hat{\sigma}_1$ is symmetric, V continuous and V coercive so that it is topologically equivalent to the V inner product.

To carry out the proof that $\mathcal{A}$ is a generator of a C_0 semigroup, we again use the Lumer-Phillips theorem. To make the arguments we need some minimal assumptions on the kernels $\tilde{g}$. In particular, we assume that $\tilde{g} > 0$ and $\dot{\tilde{g}} \geq 0$. We note that since $\tilde{g}(s) = \tilde{g}(-s)$, we have that the Debye polarization model (1.15), for example, satisfies the positivity assumption on $\tilde{g}$ as well as the monotonicity assumption $\dot{\tilde{g}} \geq 0$.

Since X_1 is a Hilbert space, it suffices to argue that for some λ_0, $\mathcal{A} - \lambda_0 I$ is dissipative in X_1 and $\mathcal{R}(\lambda I - \mathcal{A}) = X_1$ for some $\lambda > 0$, where $\mathcal{R}(\lambda I - \mathcal{A})$ is the range of $\lambda I - \mathcal{A}$. We first argue dissipativeness.

Let $\Phi = (\phi, \psi, \eta) \in \mathcal{D}(\mathcal{A})$. Then

$$
\begin{aligned}
\langle \mathcal{A}\Phi, \Phi \rangle_{X_1} &= \langle \psi, \phi \rangle_{V_1} + \langle \hat{A}\phi + B\psi + \hat{K}\eta, \psi \rangle_H + \langle \psi + C\eta, \eta \rangle_W \\
&= \langle \psi, \phi \rangle_{V_1} + \langle \hat{A}\phi + B\psi, \psi \rangle_{V^*, V} + \langle \hat{K}\eta, \psi \rangle_H + \langle \psi + C\eta, \eta \rangle_W \\
&= \hat{\sigma}_1(\psi, \phi) - \hat{\sigma}_1(\phi, \psi) + \langle B\psi, \psi \rangle_H + \langle \hat{K}\eta, \psi \rangle_H + \langle \psi + C\eta, \eta \rangle \\
&= -\langle \gamma\psi, \psi \rangle_H - c|\psi(0)|^2 + \langle \hat{K}\eta, \psi \rangle_H + \langle \psi + C\eta, \eta \rangle_W \\
&\le |\gamma|_\infty |\psi|_H^2 + |\langle \hat{K}\eta, \psi \rangle_H| + |\langle \psi + C\eta, \eta \rangle_W|.
\end{aligned}
\tag{3.2}
$$

We consider estimates for the last two terms in (3.27) separately. From (3.25) we have

$$
\begin{aligned}
|\langle \hat{K}\eta, \psi \rangle_H| &= \left| \int_{-r}^0 \tilde{g}(\theta) \langle \eta(\theta), \psi \rangle_H d\theta \right| \\
&\le \int_{-r}^0 \tilde{g}(\theta) |\eta(\theta)|_H |\psi|_H d\theta \\
&\le \frac{1}{2} \int_{-r}^0 \tilde{g}(\theta) \left\{ |\eta(\theta)|_H^2 + |\psi|_H^2 \right\} d\theta \\
&\le k_1 |\eta|_W^2 + k_2 |\psi|_H^2.
\end{aligned}
$$

Moreover,

$$
\begin{aligned}
|\langle \psi, \eta \rangle_W| &\le \int_{-r}^0 \tilde{g}(\theta) |\langle \psi, \eta(\theta) \rangle_H| d\theta \\
&\le \int_{-r}^0 \tilde{g}(\theta) \left\{ \frac{1}{2}|\psi|_H^2 + \frac{1}{2}|\eta(\theta)|_H^2 \right\} d\theta \\
&\le k_3 |\psi|_H^2 + k_4 |\eta|_W^2.
\end{aligned}
$$

Finally, since $\tilde{g} \ge 0, \dot{\tilde{g}} \ge 0$, and $\eta \in \mathcal{D}(C)$ requires $\eta(0) = 0$, we may argue

$$\langle C\eta, \eta \rangle_W = \int_{-r}^{0} \tilde{g}(\theta) \langle C\eta(\theta), \eta(\theta) \rangle_H d\theta$$

$$= \int_{-r}^{0} \tilde{g}(\theta) \frac{d}{d\theta} \frac{1}{2} |\eta(\theta)|_H^2 d\theta$$

$$= \int_{-r}^{0} \frac{d}{d\theta} \left(\tilde{g}(\theta) \frac{1}{2} |\eta(\theta)|_H^2 \right) d\theta - \int_{-r}^{0} \frac{1}{2} (\dot{\tilde{g}}(\theta) |\eta(\theta)|_H^2) d\theta$$

$$= \tilde{g}(0) \frac{1}{2} |\eta(0)|_H^2 - \tilde{g}(-r) \frac{1}{2} |\eta(-r)|_H^2 - \frac{1}{2} \int_{-r}^{0} \dot{\tilde{g}}(\theta) |\eta(\theta)|_H^2 d\theta$$

$$\leq 0.$$

By combining these estimates with (3.27), we obtain

$$\langle \mathcal{A}\Phi, \Phi \rangle_{X_1} \leq |\gamma|_\infty |\psi|_H^2 + (k_1 + k_4)|\eta|_W^2 + (k_2 + k_3)|\psi|_H^2$$
$$\leq \lambda_0 |\Phi|_{X_1}^2,$$

for $\Phi \in \mathcal{D}(\mathcal{A})$ and this yields the desired dissipativeness in X_1.

We next consider the range statement for which we must argue there exists some $\lambda > 0$ such that for any given $\Psi = (\mu, \nu, \xi)$ in X, there exists Φ in $\mathcal{D}(\mathcal{A})$ satisfying

$$(\lambda I - \mathcal{A})\Phi = \Psi$$

Given the definition of $\mathcal{A}$, this is equivalent to the system

$$\lambda\phi - \psi = \mu,$$
$$-\hat{A}\phi + (\lambda - B)\psi - \hat{K}\eta = \nu, \qquad (3.28)$$
$$-\psi + (\lambda - C)\eta = \xi$$

for $(\phi, \psi, \eta) \in \mathcal{D}(\mathcal{A})$, $(\mu, \nu, \xi) \in X = V \times H \times W$. The first equation is the same as $\psi = \lambda\phi - \mu$ while the third can be written as $\eta = (\lambda - C)^{-1}(\xi + \psi) = (\lambda - C)^{-1}(\xi + \lambda\phi - \mu)$. These two equations can be substituted in the second to obtain an equation for ϕ. If this equation can be solved for $\phi \in V$, then the first and third can be solved for ψ and η, respectively. The equation for ϕ that must be solved is given by

$$-\hat{A}\phi + (\lambda - B)(\lambda\phi - \mu) - \hat{K}(\lambda - C)^{-1}(\xi + \lambda\phi - \mu) = \nu$$

or

$$\left[\lambda^2 - \lambda B - \hat{A} - \hat{K}(\lambda - C)^{-1}\lambda \right] \phi = (\lambda - B)\mu + \nu + \hat{K}(\lambda - C)^{-1}(\xi - \mu).$$
$$(3.29)$$

If we can invert (3.29) for $\phi \in V$, then $\psi = \lambda\phi - \mu$ is in V, $\eta = (\lambda - C)^{-1}[\xi + \lambda\phi - \mu]$ is in $\mathcal{D}(C) \subset W$, and

$$
\begin{aligned}
\hat{A}\phi + B\psi &= \lambda^2\phi - \lambda\mu - \nu - \hat{K}(\lambda - C)^{-1}(\xi + \lambda\phi - \mu) \\
&= \lambda\psi - \nu - \hat{K}\eta
\end{aligned}
$$

is in H so that (ϕ, ψ, η) is in $\mathcal{D}(\mathcal{A})$ and solves (3.28).

Thus the range statement reduces to solving (3.29) for $\phi \in V$. This in turn reduces to invertibility of the operator $\lambda^2 - \lambda B - \hat{A} - \hat{K}(\lambda - C)^{-1}\lambda$.

We first observe that $(\lambda - C)^{-1} = (1 - e^{\lambda\theta})/\lambda$ since $(\lambda - C)(1 - e^{\lambda\theta}) = \lambda$ while $\eta(\theta) = \frac{1 - e^{\lambda\theta}}{\lambda}[\xi + \lambda\phi - \mu]$ satisfies $\eta(0) = 0$ and hence is in $\mathcal{D}(C)$. Thus, for $\phi \in H$, $\hat{K}(\lambda - C)^{-1}\lambda$ satisfies

$$
\begin{aligned}
\langle \hat{K}(1 - e^{\lambda\theta})\phi, \phi \rangle_H &= \int_{-r}^{0} \tilde{g}(\theta)(1 - e^{\lambda\theta})\langle \phi, \phi \rangle_H d\theta \\
&\leq k_5 |\phi|_H^2
\end{aligned}
$$

and

$$
\begin{aligned}
\langle (\lambda^2 - \lambda B)\phi, \phi \rangle_{V^*,V} &= \langle (\lambda^2 + \lambda\gamma)\phi + \lambda c\phi(0)\delta_0, \phi \rangle_{V^*,V} \\
&= \langle (\lambda^2 + \lambda\gamma)\phi, \phi \rangle_H + \lambda c|\phi(0)|^2 \\
&\geq k_6 |\phi|_H^2 \quad \text{for } \lambda \text{ sufficiently large.}
\end{aligned}
$$

Hence for λ sufficiently large we have

$$
\begin{aligned}
\langle (\lambda^2 - \lambda B - \hat{A} - \hat{K}(\lambda - C)^{-1}\lambda)\phi, \phi \rangle_{V^*,V} \\
= \langle (\lambda^2 - \lambda B)\phi, \phi \rangle_{V^*,V} + \hat{\sigma}_1(\phi, \phi) - \langle \hat{K}(\lambda - D)^{-1}\lambda\phi, \phi \rangle_H \\
\geq k_6 |\phi|_H^2 + c_1 |\phi|_V^2 - \lambda_0 |\phi|_H^2 - k_5 |\phi|_H^2 \\
= c_1 |\phi|_V^2 + (k_6 - \lambda_0 - k_5)|\phi|_H^2.
\end{aligned}
$$

Thus if we define the sesquilinear form (this will be explained carefully in Section 5.4 below but we shall go ahead and complete the arguments here and refer readers to Sections 5.4–5.6)

$$
\sigma_\lambda(\phi, \psi) \equiv \langle (\lambda^2 - \lambda B - \hat{A} - \hat{K}(\lambda - C)^{-1}\lambda)\phi, \psi \rangle_{V^*,V},
$$

we see that for λ sufficiently large, σ_λ is V coercive and hence, by the Lax-Milgram lemma (see [W] as well as Sections 5.5 and 5.6), it is invertible. It follows immediately that (3.29) is invertible for $\phi \in$

V. This completes the arguments to prove that $\mathcal{A}$ is the infinitesimal generator of a C_0 semigroup.

Let $S(t)$ denote the semigroup generated by $\mathcal{A}$ so that solutions to (3.26) are given by

$$x(t) = S(t)x_0 + \int_0^t S(t-s)\mathcal{F}(s)ds. \tag{3.30}$$

We note that solutions are clearly continuously dependent on initial data x_0 and the nonhomogeneous perturbation $\mathcal{F}$ (this will be of interest in later discussions). The first component of $x(t)$ is a solution $E(t)$ of (1.20).

4 Adjoint Operators and Dual Spaces

4.1 Adjoint Operators

Recall that $A^* : \mathcal{D}(A^*) \subset X \to X$ is defined in Definition 3.2 and we required that A be densely defined in X. We argue that this defines A^* uniquely if $\mathcal{D}(A)$ is dense in X. To show this, suppose there exist w_1 and w_2 such that

$$\langle Ax, y \rangle = \langle x, w_1 \rangle \qquad \text{for all } x \in \mathcal{D}(A)$$

and

$$\langle Ax, y \rangle = \langle x, w_2 \rangle \qquad \text{for all } x \in \mathcal{D}(A).$$

Then,

$$\langle x, w_1 - w_2 \rangle = 0 \qquad \text{for all } x \in \mathcal{D}(A).$$

If $\mathcal{D}(A)$ is dense in X, then $w_1 - w_2$ must equal zero. (To see this take $\{x_n\}$ in $\mathcal{D}(A)$ such that $x_n \to w_1 - w_2$ and then take the limit in $\langle x_n, w_1 - w_2 \rangle = 0$ to obtain $|w_1 - w_2| = 0$.) Therefore, $w_1 = w_2$, and the adjoint of A is uniquely defined.

4.1.1 Computation of A^*: An Example from the Heat Equation

The computation of A^* can be easy or impossible. Let $X = L_2(0, l)$, then the operator A from the heat equation is defined by

$$A\varphi = (D\varphi')'$$

on

$$\mathcal{D}(A) = \{\varphi \in H^2(0, l) \mid (D\varphi')' \in L_2(0, l), \ \varphi(0) = \varphi'(l) = 0\}.$$

Note that this makes more precise the statement "D is smooth" in Section 1.3. For $\varphi \in \mathcal{D}(A)$, D real, and $\psi \in L_2(0, l) = X$, we have

$$
\begin{aligned}
\langle A\varphi, \psi \rangle &= \int_0^l (D\varphi')' \bar{\psi} \\
&= -\int_0^l D\varphi' \bar{\psi}' + D\varphi' \bar{\psi} \big|_0^l \\
&= \int_0^l \varphi (D\bar{\psi}')' - \varphi D\bar{\psi}' \big|_0^l + D\varphi' \bar{\psi} \big|_0^l \\
&= \langle \varphi, (D\bar{\psi}')' \rangle - \varphi D\bar{\psi}' \big|_0^l + D\varphi' \bar{\psi} \big|_0^l \\
&= \langle \varphi, w \rangle \qquad \text{for all } \varphi \in \mathcal{D}(A)
\end{aligned}
$$

47

if and only if $\psi'(l) = 0$ and $\psi(0) = 0$. (We must, of course, have $\psi \in H^2(0, l)$ and $(D\psi')' \in L_2(0, l)$ to carry out the operations.) Hence

$$\mathcal{D}(A^*) = \{\psi \in H^2(0, l) \mid (D\psi')' \in L_2(0, l), \ \psi(0) = \psi'(l) = 0\}.$$

and

$$A^*\psi = (D\psi')'.$$

Therefore $A = A^*$, $\mathcal{D}(A) = \mathcal{D}(A^*)$, and hence A is a self-adjoint operator.

Further examples involving computation of adjoints will be give in Section 4.6 below.

4.1.2   Self-Adjoint Operators and Dissipativeness

For self-adjoint linear operators, if $\varphi \in \mathcal{D}(A) = \mathcal{D}(A^*)$, we have

$$\mathrm{Re}\ \langle A\varphi, \varphi \rangle = \mathrm{Re}\ \langle \varphi, A^*\varphi \rangle = \mathrm{Re}\ \langle A^*\varphi, \varphi \rangle.$$

Therefore, whenever A is self-adjoint, A is dissipative if and only if A^* is dissipative. Thus, for any closed, densely-defined, dissipative self-adjoint operator in X, both A and A^* are infinitesimal generators (see Corollary 5 to the Lumer-Phillips theorem) of C_0 semigroups $T(t) \sim e^{At}$ and $S(t) \sim e^{A^*t}$. From a fundamental Hilbert space result, given below, we know that $S(t) = T(t)^*$.

Theorem 4.1 *If A is an infinitesimal generator of a C_0 semigroup $T(t)$ on a Hilbert space X, then $T(t)^* = S(t)$ is a C_0 semigroup on X with infinitesimal generator A^*.*

Homework Exercise

- Ex. 8 (Project 1): Consider $\mathcal{D}(A^*)$ in a Hilbert space X where A is densely defined in X.

    - (a) What can you say about $\mathcal{D}(A^*)$ with respect to closed, dense, nonempty, all in X, relationship between A^{-1} and $(A^*)^{-1}$ when they exist, and symmetric versus self-adjoint for the operator A?

    - (b) Reconsider all the same questions in a Banach space X (Note: In a Banach space $\mathcal{D}(A^*) \subset X^*$) after studying Sections 4.4 and 4.5 below.

Give complete examples and counter examples. Also, give complete references.

For relevant material, see [DS, HP, K, RN, Y].

4.2 Dual Spaces and Strong, Weak, and Weak* Topologies

Before discussing dissipative operators in a Banach space X, we must discuss topological dual spaces X^*. We talk about strong, weak, and weak* topologies in a Banach space X. For relevant material, see [DS, K, Y].

Let X be a Banach space. Then recall that X^*, called the *dual space* or *"topological" dual space*, is a Banach space whenever X is a normed linear space and the scalar field is complete. It was defined as the Banach space of continuous conjugate or anti-linear functionals on X, where a conjugate linear functional satisfies $f(\alpha x + \beta y) = \bar{\alpha} f(x) + \bar{\beta} f(y)$ for all $x, y \in X$ and scalars α, β. Again recall that elements of X^* are denoted by x^* where $|x^*|_* = \sup\limits_{|x| \leq 1} |x^*(x)|$, with $|\cdot|_*$ denoting the norm in X^*. This of course gives rise to a strong or normed topology of X^*.

If X is a complex Hilbert space, then we have the Riesz theorem.

Theorem 4.2 *(Riesz theorem)[K, p.253],[Sh, p.16] If X is a Hilbert space, then for every $f \in X^*$, there is a corresponding $\tilde{f} \in X$ such that*

$$f(x) = \langle \tilde{f}, x \rangle$$

and

$$f(\alpha x + \beta y) = \langle \tilde{f}, \alpha x + \beta y \rangle = \bar{\alpha}\langle \tilde{f}, x \rangle + \bar{\beta}\langle \tilde{f}, y \rangle$$

for all $x \in X$.

One can use this theorem to prove that $X \cong X^*$. The reader should compare this version of the Riesz theorem with the more usual version of the Riesz theorem when X^* is defined as linear (as opposed to conjugate linear or anti-linear) continuous functionals (e.g., see [Y, p.90],[NS, p.345]). Of course, if the scalar field $\mathbb{F}$ is the real numbers $\mathbb{R}^1$, then these are exactly the same.

Definition 4.1 *Weak convergence* of a sequence $\{x_n\} \subset X$, $x_n \rightharpoonup x$, is defined by

$$x_n \rightharpoonup x \text{ if and only if } x^*(x_n) \to x^*(x) \text{ for all } x^* \in X^*.$$

We write $x = \text{wlim } x_n$.

Definition 4.2 X is said to be *weakly sequentially complete* if and only if every weakly Cauchy sequence ($\{x_n\}$ such that $x_n - x_m \rightharpoonup 0$ as $n, m \to \infty$) in X converges weakly to an element $x \in X$.

Result 4.1 *There are a number of useful results related to weak convergence.*

1. *Weak limits are unique.*

   To see this, suppose $x_n \rightharpoonup x$ and $x_n \rightharpoonup y$, then $x^*(x) = x^*(y)$, i.e, $x^*(x - y) = 0$ for all $x^* \in X^*$. However, the Hahn-Banach theorem says if $z \neq 0$, then there exists an $x^* \in X^*$ such that $x^*(z) \neq 0$. Therefore, $x = y$.

2. *(a) $x_n \to x$ implies $x_n \rightharpoonup x$, but the converse is not true.*

   (b) $x_n \rightharpoonup x$ implies $\{x_n\}$ is bounded in X and $|w \lim x_n| = |x| \leq \lim \inf |x_n|$. Note: This says that the norm $|\cdot|$ in X is weakly lower semi-continuous. Recall if f is continuous, then $|f(x) - f(x_0)| < \epsilon$ or $f(x_0) - \epsilon < f(x) < f(x_0) + \epsilon$ for x near x_0. Lower semi-continuous means we just have $f(x_0) - \epsilon < f(x)$ for x sufficiently near x_0 or equivalently $\lim \inf_{x \to x_0} f(x) \geq (x_0)$. This has important applications in optimization (see the remark below).

   (c) $x_n \to x$ if and only if $x^*(x_n)$ converges uniformly for $|x^*|_* \leq 1$.

Remark: Application of Semi-continuity in Optimization

Weak lower semi-continuity plays a fundamental role in many optimization settings. Suppose we are in a Hilbert space X and we want to minimize a function $J(x)$ over $K \subset X$. Further suppose we have a minimizing sequence $\{x_n\}$ such that $J(x_n) \searrow \inf_{x \in K} J(x)$ and $\{x_n\}$ is bounded, i.e., $|x_n| \leq M$. As we shall see below, bounded sequences possess weakly convergent subsequences. Therefore we can choose a subsequence converging weakly, i.e., $x_{n_k} \rightharpoonup \tilde{x}$. If J is weakly lower semi-continuous, then we have $J(\text{wlim } x_{n_k}) \leq \lim \inf J(x_{n_k})$ or $J(\tilde{x}) \leq \inf_{x \in K} J(x)$. Whenever K is weakly sequentially closed, this yields existence of a minimum (not just an infimum) when minimizing a weakly lower semi-continuous J. These facts are often used in constructive arguments to develop computational algorithms.

Definition 4.3 Weak convergence generates a topology in X (one can use generalized sequences or nets to generate this topology) called the *weak topology of X* or *X^* topology of X*.

Note: Weakly closed implies strongly closed, but not conversely in general. However, we do have an important case for which weak and strong closure are equivalent.

Theorem 4.3 *(Mazur - 1933) A convex set $E \subset X$ is X^* closed if and only if X is closed.*

Thus, for convex sets, a set is weakly closed if and only if it is strongly closed. See [DS] for relevant information.

Definition 4.4 Since X^* is a Banach space, we can define its dual, X^{**}. There is a natural embedding $X \hookrightarrow X^{**}$ defined by $J : X \to X^{**}$ where

$$J(x)(x^*) = x^*(x)$$

for all $x^* \in X^*$.

In general, $J(X) \subset X^{**}$, but $J(X) \neq X^{**}$.

Definition 4.5 If $J(X) = X^{**}$, then X is called a *reflexive Banach space*.

Result 4.2 *The following elementary results can be argued (again see [DS, K, T] for details and proofs).*

1. *Hilbert spaces are reflexive.*

2. *X is a reflexive Banach space implies X is weakly sequentially complete.*

3. *X is a reflexive Banach space if and only if $\{x \in X | \, |x| \leq 1\}$ is compact in the weak topology. Therefore, in a reflexive Banach space, bounded sets are weakly, sequentially pre-compact.*

4. *Bounded sets in a Hilbert space are weakly, sequentially pre-compact. (We have this from Results 1 and 3.)*

5. *Let X be a Hilbert space. Then*

$$x_n \rightharpoonup x \text{ and } |x_n| \to |x| \text{ implies } x_n \to x.$$

Conversely,

$$x_n \rightharpoonup x \text{ and } x_n \to x \text{ implies } |x_n| \to |x|.$$

Therefore,

$$x_n \rightharpoonup x \text{ implies } x_n \to x \text{ if and only if } |x_n| \to |x|.$$

This is easy to see by using

$$|x_n - x|^2 = \langle x_n - x, x_n - x \rangle = |x_n|^2 - \langle x_n, x \rangle - \langle x, x_n \rangle + |x|^2.$$

Definition 4.6 For any Banach space X, X^* is itself a Banach space with norm $|\cdot|_*$; therefore, the weak topology of X^* can be defined and is called the X^{**} *topology*.

We can now consider weak convergence in X^* which can be characterized by

$$x_n^* \rightharpoonup x^* \in X^* \text{ if and only if } x^{**}(x_n^*) \to x^{**}(x^*) \text{ for all } x^{**} \in X^{**}.$$

Definition 4.7 We can put an even weaker topology on X^*. We can define a *weak* convergence* of $\{x_n^*\} \subset X^*$, denoted as $x_n^* \overset{w*}{\rightharpoonup} x^*$, by

$$x_n^* \overset{w*}{\rightharpoonup} x^* \equiv x_n^*(x) \to x^*(x) \quad \text{for all } x \in X.$$

(i.e., the pointwise convergence of conjugate linear functionals.) Note that this is in general a weaker convergence than weak convergence since $J(X) \subset X^{**}$ with $J(X) \neq X^{**}$ in general.

Result 4.3 *We have the following results.*

1. $x_n^* \overset{w*}{\rightharpoonup} x^*$ *implies* $|x_n^*|_* \leq M$.

2. $x_n^* \to x^*$ *in* X^* *implies* $x_n^* \overset{w*}{\rightharpoonup} x^*$ *in* X^* *(i.e., strong convergence implies weak* convergence).*

   The converse does not hold.

3. *If X is a Banach space, then X^* is weak* sequentially complete and the norm $|\cdot|_*$ is weak* lower semi-continuous. In other words,*

$$|w^* \lim x_n^*|_* \leq \lim \inf |x_n^*|_*.$$

4. $x_n^* \rightharpoonup x^*$ *in X^* implies $x_n^* \overset{w^*}{\longrightarrow} x^*$ in X^* (i.e., weak convergence implies weak* convergence).*

The weak* convergence in X^* can be used to define a topology for X^*, called the X topology of X^*. In general, the X topology of X^* is weaker than the X^{**} topology of X^*. In other words, in X^* we have

strong convergence $\Rightarrow$ weak convergence $\Rightarrow$ weak* convergence.

However, the converses do not hold.

4.2.1 Summary of Topologies on X and X^*

X	X^*
strong	strong
weak	weak
(X^* topology of X)	(X^{**} topology of X^*)
–	weak *
	(X topology of X^*)

Important Summary Note:

For any Banach space Y, we can talk about weak convergence, but we can only talk about weak* convergence in the case that $Y = X^* = $ the dual of some space X !!

A very important theorem can be found in [DS, p. 424], [Royden, p. 202].

Theorem 4.4 *(Alaoglu's theorem) Let X be a Banach space. Then the unit ball (closed unit sphere) in X^* is compact in the weak* topology (i.e., the X topology of X^*).*

Corollary 4.1 *Norm $(|\cdot|_*)$ bounded sets in X^* are weak* sequentially compact (i.e., every sequence in the set has a subsequence weak* converging to a point in X^*).*

Result 4.4 *We can summarize additional findings that are conse-quences of the above results.*

1. *Let $J : X \to X^{**}$ be the natural embedding so that for a reflex-ive Banach space, $J(X) = X^{**}$. Hence we have that the weak and weak* topologies of X^* are the same. To see this we argue $x^{**}(x_n^*) = J(x)(x_n^*) = x_n^*(x)$ and hence*

$$x^{**}(x_n^*) \to x^{**}(x^*)$$

is weak sense convergence with

$$x^{**}(x^*) = J(x)(x^*) = x^*(x).$$

Since $x_n^*(x) \to x^*(x)$ is weak* convergence, these are thus equiv-alent.

Conversely, if X is not reflexive, then $J(X) \neq X^{**}$ and the X topology of X^* is weaker than the X^{**} topology of X^*. Thus in a reflexive Banach space (which includes Hilbert spaces), bounded sets are weakly sequentially compact. *(This is a special case of Alaoglu's theorem.)*

2. *A Banach space is reflexive if and only if the closed unit ball is compact in weak topology.*

4.3 Examples of Spaces and Their Duals

- We first consider $L_p(0,1)$ and recall [Royden, p. 130-132]

$$L_p^*(0,1) = L_q(0,1), \quad 1 \leq p < \infty, \quad \frac{1}{p} + \frac{1}{q} = 1.$$

In other words, L_∞ is the dual of a space, $L_1^* = L_\infty$; however, we do not have a satisfactory representation for the dual L_∞^* of L_∞. In other words, we cannot write L_1 as the dual of L_∞ or any other known space. As a result, we can consider the weak* topology in L_p for $1 < p \leq \infty$, but we cannot talk about the weak* topology in L_1.

- $L_p(0,T;X)$ are important spaces because they are very useful in studying the system $\dot{x}(t) = Ax(t) + F(t)$, $0 < t < T$, in X. We have (to be discussed later—see Chapter 7 and Theorem 7.1)

$$L_p(0,T;X)^* \cong L_q(0,T;X^*), \quad 1 \leq p < \infty.$$

See [E] for relevant details on Banach space valued functions of time on intervals $(0, T)$.

- The Banach space of (bounded) continuous functions on $[a, b]$ is denoted by $C[a, b]$. It has a norm defined by $|\varphi| = \sup_{\xi \in [a,b]} |\varphi(\xi)|$ for $\varphi \in C[a, b]$. A similar definition holds for $C(Q)$ for Q a general metric space (see [DS, p. 261]) where the notation $C_B(Q)$ is sometimes used for the bounded continuous functions (if Q is compact then of course any continuous function is bounded).

- The Banach space of regular bounded additive set functions is denoted by rba$[a, b]$. Suppose μ is a regular set function on Borel subsets of $[a, b]$. Then by definition of *regular*, given a set $F \subset [a, b]$ and $\epsilon > 0$, there exists a closed set E and an open set O such that $E \subset F \subset \bar{O}$ and $\mu(\bar{O} - E) < \epsilon$. Additive means that for any two Borel subsets A and B contained in $[a, b]$ such that $A \cap B = \emptyset$, we have $\mu(A \cup B) = \mu(A) + \mu(B)$. Again we note that one can extend this definition to define rba(Q) for a metric space Q-see [DS, p. 261].

- The Banach space of normalized functions of bounded variation on an interval $[a, b]$ is denoted by NBV$[a, b]$. The space BV$[a, b]$ has a norm $|f| = |f(a^+)| + \text{var}(f)$ where $\text{var}(f)$ is the total variation

$$\text{var}(f) = \sup_{\pi[a,b]} \sum_{i=1}^{N} |f(x_i) - f(x_{i-1})|,$$

where $\pi[a, b]$ denotes the partitions of $[a, b]$ and $f(a^+) = \lim_{\{x \to a, x > a\}} f(x)$. Note that $\text{var}(f)$ provides a semi-norm ($\text{var}(f) = 0$ does not imply $f = 0$) on BV. Moreover BV can be written as BV$_o \oplus N$ where BV$_o = \{f \in \text{BV} | f(a^+) = 0\}$ and N is the 1-D space of constant functions.

NBV$[a, b]$ is normalized by making f right-continuous at the interior points and $f(a^+) = 0$ with $|f| = \text{var}(f)$.

Then we have the following theorem from [DS, Chapter IV, p. 262,337].

Theorem 4.5 *The following equivalences hold:*

$$C^*[a, b] \cong rba[a, b] \cong NBV[a, b]$$

where NBV[a, b] stands for normalized bounded variation functions and rba[a, b] stands for regular bounded additive set functions. More generally, $C_B^*(Q) \cong rba(Q)$ for a metric space Q.

Therefore we can consider the weak* topology for NBV[a, b]; however, C[a, b] is not the dual of another space, and specifically NBV[a, b]* ≠ C[a, b] (i.e., we do not have satisfactory characterizations: there is no known space X such C[a, b] = X*). Thus we cannot talk about the weak* topology in C[a, b] (and of course C[a, b] is not a reflexive Banach space).

Discussion of NBV[a, b]:

Suppose $f \in$ NBV[a, b]; then we can write [Ru1, p. 101], [Ru2, p. 163]

$$
\begin{aligned}
f(x) &= f(a) + p_f(x) - n_f(x) \\
v(x) &= p_f(x) + n_f(x),
\end{aligned}
$$

where p_f, n_f and v are nondecreasing monotone functions defined by

$$
p_f(x) = \sup_{\pi[a,x]} \sum_{i=1}^{N} (f(x_i) - f(x_{i-1}))^+,
$$

$$
n_f(x) = \sup_{\pi[a,x]} \sum_{i=1}^{N} |(f(x_i) - f(x_{i-1}))^-|, \quad \text{and}
$$

$$
v(x) = \sup_{\pi[a,x]} \sum_{i=1}^{N} |f(x_i) - f(x_{i-1})|,
$$

where $(\cdot)^+$ and $(\cdot)^-$ are the positive and negative parts of $(\cdot)$, respectively. Using the monotone functions p_f and n_f, one can generate (positive) Lebesgue-Stieltjes measures μ_{p_f} and μ_{n_f} [DS, Hal, HeSt, McS, Ru1, Ru2, Smir] and subsequently a signed measure

$$
\mu_f = \mu_{p_f} - \mu_{n_f},
$$

where μ_f is regular. From the above theorem we have $C^*[a, b] \cong$ rba(a, b), which means there is a one-to-one correspondence

$$
x^* \leftrightarrow \mu,
$$

given by

$$x^*(\varphi) = \int \varphi d\mu.$$

We also have $C^*[a, b] \cong \mathrm{NBV}[a, b]$, which means we have

$$x^* \leftrightarrow \mu_f,$$

given by

$$x^*(\varphi) = \int \varphi d\mu_f = \int \varphi df,$$

where the integrals are Lebesgue-Stieltjes integrals.

- We also need to consider Sobolev spaces, $W^{k,p}(\Omega)$. We note that $W^{k,p}(\Omega)$ is reflexive for $1 < p < \infty$ and, moreover, $H^k(\Omega) = W^{k,2}(\Omega)$ is a Hilbert space. We can define $W_0^{k,p}(\Omega)$ as Sobolev spaces with compact support ("functions vanish at the boundary"), or we can think of $W_0^{k,p}(\Omega)$ as the closure of $C_0^\infty(\Omega)$ in $W^{k,p}$. We can consider their dual spaces and find

$$W_0^{k,p}(\Omega)^* \cong W^{-k,p}(\Omega) \quad 1 < p < \infty, \quad \frac{1}{p} + \frac{1}{q} = 1,$$

where the negative indices indicate generalized derivatives (for relevant material, see [Adams, DS]).

4.4 Return to Dissipativeness for General Banach Spaces

Definition 4.8 Let X be a Banach space and X^* be its conjugate dual. Then

$$x^*(x) = \langle x^*, x \rangle_{X^*, X}$$

is called the *duality product*. This is sometimes written as simply $\langle x^*, x \rangle$ since it is a generalization of the inner product in a Hilbert space. Indeed, if X is a Hilbert space with $X^* \cong X$, then $\langle x^*, x \rangle_{X^*, X}$ reduces to the inner product $\langle x^*, x \rangle_X$ in X.

Definition 4.9 For each x in a Banach space X, we define the *duality set* $F(x) \subset X^*$ by

$$F(x) = \{ x^* \in X^* | \langle x^*, x \rangle = |x|_X^2 = |x^*|_{X^*}^2 \}.$$

We know $F(x) \neq \emptyset$ for each $x \in X$ by the Hahn-Banach theorem.

Result 4.5 *If X is a Hilbert space, one can show $F(x) = \{x\}$ under the usual isomorphism $X \simeq X^*$. (This is the Duality Set Lemma, Lemma 3.1, page 31.)*

Definition 4.10 Let $A : \mathcal{D}(A) \subset X \to X$ for a Banach space X. Then A is called *dissipative* if for <u>every</u> $x \in \mathcal{D}(A)$ there exists <u>some</u> $x^* \in F(x)$ such that

$$\mathrm{Re}\langle x^*, Ax \rangle \leq 0.$$

Result 4.6 The theorems stated previously for Hilbert spaces are still true in a Banach space with the definition of dissipativeness given above. *For discussions of the Lumer-Phillips characterizations, and related results in a Banach space, see [Pa, p. 13-16]. We shall not present these important results here since we do not use them in subsequent discussions in this book.*

4.5   More on Adjoint Operators

We recall that for $T \in \mathcal{L}(X, Y)$ where X, Y are normed linear spaces, then $T^* \in \mathcal{L}(Y^*, X^*)$ is defined by $T^*(y^*)(x) = y^*(Tx)$ for all $x \in X$. In general, we do not need to have $T \in \mathcal{L}(X, Y)$ for T^* to be defined; however, if we do have $T \in \mathcal{L}(X, Y)$, then $T^* \in \mathcal{L}(Y^*, X^*)$. We generalize our concept of adjoint operators in a Hilbert space (see Definition 3.2) to unbounded operators between normed linear spaces.

   Let X and Y be normed linear spaces with X^* and Y^* as their conjugate duals, respectively. For $\mathcal{D}(A)$ dense in X, suppose we have the operator $A : \mathcal{D}(A) \subset X \to Y$. Then we can define the adjoint operator $A^* : \mathcal{D}(A^*) \subset Y^* \to X^*$ with

$$\mathcal{D}(A^*) = \{y^* \in Y^* \mid x \to y^*(Ax) \text{ is continuous on } \mathcal{D}(A)\}.$$

Given $y^* \in Y^*$, we see that $y^* \in \mathcal{D}(A^*)$ if and only if there exists $x^* \in X^*$ such that $y^*(Ax) = x^*(x)$ for all $x \in \mathcal{D}(A)$. Therefore, we define A^* by

$$A^* y^* = x^*.$$

In other words,

$$(A^* y^*)(x) = x^*(x) = y^*(Ax).$$

4.5.1   Adjoint Operator in a Hilbert Space

In a Hilbert space, $X^* \sim X$ and $Y^* \sim Y$; therefore,

$$\langle Ax, y \rangle_Y = \langle x, A^* y \rangle_X.$$

4.5.2   Special Case

Theorem 4.6 *Let X and Y be reflexive Banach spaces. If T is a closed linear operator $T : X \to Y$ with $\mathcal{D}(T)$ dense in X, then $\mathcal{D}(T^*)$ is dense in $Y^*(\sim Y)$ and $T^{**} = T$.*

Note: This does not say anything about the relationship between $\mathcal{D}(T)$ and $\mathcal{D}(T^*)$. As an exercise, the reader should return to Exercise 8(b) of Section 4.1 and reconsider the questions posed there in the context of adjoint operators and their domains in Banach spaces.

4.6   Examples of Computing Adjoints

In this series of examples, we illustrate how the domain, and in particular, boundary conditions, can affect the definition of an operator. In this case we study operators which are all essentially the operations of second-order differentiation, i.e., $A\varphi = \varphi''$, arising in the heat equation operator of Example 1. As we shall see, the boundary conditions affect dramatically the nature of the adjoint A^*.

We define $H_0^2(0, 1) = \{\varphi \in H^2(0, 1) | \varphi(0) = \varphi(1) = \varphi'(0) = \varphi'(1) = 0\}$.

1. Let $H = L_2(0, 1)$ and $V = H_0^2(0, 1)$. We define the operator A by

$$A : \mathcal{D}(A) = V \subset H \to H$$

by

$$A\varphi = \varphi''.$$

In order to define the adjoint operator A^*, we need to find, for each ψ, an element w such that

$$\langle A\varphi, \psi \rangle = \langle \varphi, w \rangle \quad \text{for all } \varphi \in H_0^2(0, 1).$$

Therefore, we may carry out the calculations

$$
\begin{aligned}
\langle A\varphi, \psi \rangle &= \int_0^1 \varphi''\psi \\
&= -\int_0^1 \varphi'\psi' + \varphi'\psi|_0^1 \\
&= \int_0^1 \varphi\psi'' - \varphi\psi'|_0^1 + \varphi'\psi|_0^1 \\
&= \int_0^1 \varphi\psi'' + 0 + 0 \\
&= \langle \varphi, \psi'' \rangle
\end{aligned}
$$

which hold if and only if $\psi \in H^2$.

Therefore

$$
\mathcal{D}(A^*) = H^2(0,1)
$$

with

$$
w = A^*\psi = \psi''.
$$

Observe that $\mathcal{D}(A^*) \neq \mathcal{D}(A)$; therefore, A is not self-adjoint. However, $A^* = A$ on their common domain H_0^2. Thus we see that there are boundary conditions which immediately render the heat operator non-self-adjoint.

2. Let $H = L_2(0,1)$, $V = H_0^2(0,1)$ and recall that $V^* = H^{-2}(0,1) \equiv \{\psi | x \to \int_0^x \int_0^s \psi(t)dtds$ is in $L_2(0,1)\}$ (see [Adams]). We define the operator $\tilde{A}$ by

$$
\tilde{A} : H \to V^*
$$

by

$$
(\tilde{A}\varphi)\psi = \langle \varphi, \psi'' \rangle_H = \langle \varphi, \psi'' \rangle_{L_2}
$$

for $\varphi \in H$ and $\psi \in V$. We claim that $\tilde{A} \in \mathcal{L}(H, V^*)$ and this shows that differentiation can be bounded (continuous) if the spaces are chosen in a certain way (it is a standard exercise that differentiation from $L_2 \to L_2$ is not bounded). To verify our claim (since $\tilde{A}$ is clearly linear), we show $|\tilde{A}\varphi|_{V^*} \leq k|\varphi|_H$. We have

$$
\begin{aligned}
|(\tilde{A}\varphi)(\psi)| &= |\langle \varphi, \psi'' \rangle| \\
&\leq |\varphi|_{L_2}|\psi''|_{L_2} \\
&\leq |\varphi|_H|\psi|_V.
\end{aligned}
$$

Therefore

$$
\frac{|(\tilde{A}\varphi)(\psi)|}{|\psi|_V} \leq |\varphi|_H.
$$

However

$$|\tilde{A}\varphi|_{V^*} = \sup_{\psi \neq 0} \frac{|(\tilde{A}\varphi)(\psi)|}{|\psi|_V}.$$

Thus

$$|\tilde{A}\varphi|_{V^*} \leq |\varphi|_H$$

which implies that $\tilde{A} \in \mathcal{L}(H, V^*)$. From this we know $\tilde{A}^* \in \mathcal{L}(V, H)$. Next we seek to find $\tilde{A}^*$ by using the definition

$$(\tilde{A}^*\psi)(\varphi) = (\tilde{A}\varphi)(\psi) = \langle \varphi, \psi'' \rangle$$

for $\varphi \in H$. Thus $\tilde{A}^*$ is defined on all of V by

$$\tilde{A}^*\psi = \psi''.$$

As in the previous example, $\tilde{A}$ is not self-adjoint.

3. Let $H = L_2(0,1)$, $V = H_0^1(0,1) = \{\varphi \in H^1(0,1)|\varphi(0) = \varphi(1) = 0\}$ and define the operator A_1 by

$$A_1 : \mathcal{D}(A_1) \subset H \to H$$

where

$$\mathcal{D}(A_1) = H^2(0,1) \cap H_0^1(0,1)$$

with

$$A_1\varphi = \varphi''.$$

We can argue $A_1 \notin \mathcal{L}(H)$ as in a previous assignment (see Ex. 2, Section 1.3). In order to find the adjoint operator A_1^*, we want for each ψ to find a w such that

$$\langle A_1\varphi, \psi \rangle = \langle \varphi, w \rangle \quad \text{for all } \varphi \in \mathcal{D}(A_1).$$

However

$$
\begin{aligned}
\langle A_1\varphi, \psi \rangle &= \int_0^1 \varphi''\psi \\
&= \int_0^1 \varphi\psi'' - \varphi\psi'|_0^1 + \varphi'\psi|_0^1 \\
&= \int_0^1 \varphi\psi'' + 0 + \varphi'\psi|_0^1 \\
&= \int_0^1 \varphi\psi''
\end{aligned}
$$

if and only if $\psi \in H^2 \cap H_0^1$. Hence

$$A_1^*\psi = \psi''$$

with

$$\mathcal{D}(A_1^*) = H^2(0,1) \cap H_0^1(0,1) = \mathcal{D}(A_1).$$

Thus A_1 is self-adjoint; however, it is not a bounded linear operator on H.

4. Let $V = H_0^1(0,1)$. Then $V^* = H^{-1}(0,1)$. We can define the operator $\tilde{A}_1$ by

$$\tilde{A}_1 : V \to V^*$$

with

$$(\tilde{A}_1\varphi)(\psi) = -\langle \varphi', \psi' \rangle_{L_2}.$$

We argue $\tilde{A}_1 \in \mathcal{L}(V, V^*)$. The operator is clearly linear so that to see this we must establish $|\tilde{A}_1\varphi|_{V^*} \le k|\varphi|_V$ for some k. We find

$$
\begin{aligned}
|(\tilde{A}_1\varphi)(\psi)| &= |\langle \varphi', \psi' \rangle| \\
&\le |\varphi'|_{L^2}|\psi'|_{L^2} \\
&\le |\varphi|_V |\psi|_V,
\end{aligned}
$$

and hence for all ψ

$$\frac{|(\tilde{A}_1\varphi)(\psi)|}{|\psi|_V} \le |\varphi|_V.$$

However

$$|\tilde{A}_1\varphi|_{V^*} = \sup_{\psi \ne 0} \frac{|(\tilde{A}_1\varphi)(\psi)|}{|\psi|_V}.$$

Therefore

$$|\tilde{A}_1\varphi|_{V^*} \le |\varphi|_V$$

which yields that $\tilde{A}_1 \in \mathcal{L}(V, V^*)$. We know that $\tilde{A}_1^* \in \mathcal{L}(V, V^*)$. Next we would like to find $\tilde{A}_1^*$ which must satisfy

$$(\tilde{A}_1^*\psi)(\varphi) = (\tilde{A}_1\varphi)(\psi) = -\langle \varphi', \psi' \rangle$$

for $\varphi \in \mathcal{D}(\tilde{A}_1)$. Hence $\tilde{A}_1^*$ is defined on all of V by

$$(\tilde{A}_1^*\psi)\varphi = -\langle \varphi', \psi' \rangle.$$

Thus we find that $\tilde{A}_1$ is a bounded self-adjoint linear operator.

Remark 1 *We summarize the calculations above by noting that we see that the choice of boundary conditions and the spaces in which the operators are defined determine the properties of the associated operator formulation. We shall see further evidence of this in Sections 6.1 and 6.2.*

5 Gelfand Triple, Sesquilinear Forms, and Lax-Milgram

We recall in our use of the Lumer-Phillips generation theorem (Theorem 3.4) with several examples a major task was verifying the range statement which is equivalent to solving equations of the form $(\lambda_0 - A)\phi = \psi$ for any given $\psi \in X$. In this section we introduce and discuss fundamental mathematical concepts, sesquilinear forms, that will provide an important tool in treating succinctly such range statements. We then will discuss several forms of the celebrated Lax-Milgram theorem that readily yield the desired range results needed in the Lumer-Phillips generation theorem for numerous examples. To motivate and illustrate our discussions, we introduce another example which is ubiquitous in structural applications and is also ideal for use in illustrating our weak formulations (as opposed to the semigroup formulations) of equations that follow in subsequent sections of this book.

5.1 Example 6: The Cantilever Beam

We consider the beam equation given by

$$\rho\frac{\partial^2 y}{\partial t^2} + \gamma\frac{\partial y}{\partial t} + \frac{\partial^2}{\partial \xi^2}M = f(t,\xi) \qquad (5.1)$$

where the term $\gamma\frac{\partial y}{\partial t}$ represents the external damping (the air or viscous damping), M is the internal moment, ρ is the mass density, and f is the external force applied. For a development of this equation from basic principles, see [BSW, BT]. Before discussing the equation further, we discuss boundary and initial conditions.

Boundary Conditions We consider a beam with one end fixed and one end free. The general boundary conditions for such a structural configuration are given by

Fixed end:

$$
\begin{aligned}
y(t,0) &= 0 \quad \text{for zero displacement,} \\
\tfrac{\partial y}{\partial \xi}(t,0) &= 0 \quad \text{for zero slope.}
\end{aligned}
$$

Free end:

$$
\begin{aligned}
M(t,l) &= 0 \quad \text{for zero moment,} \\
\tfrac{\partial M}{\partial \xi}(t,l) &= 0 \quad \text{for zero shear.}
\end{aligned}
$$

We observe that the shear force is represented by $S(t, \xi) = -\frac{\partial M}{\partial \xi}(t, \xi)$.

Initial Conditions We consider a beam with initial displacement and velocity given by Φ and Ψ, respectively.

$$
\begin{aligned}
y(0, \xi) &= \Phi(\xi) \\
\dot{y}(0, \xi) &= \Psi(\xi)
\end{aligned}
$$

We observe that we have an equation (5.1) given with two unknowns; however we couple these with a *constitutive relationship* given by

$$M = M(y). \tag{5.2}$$

Using Hooke's law ($\sigma = E\epsilon$ where σ is the stress, ϵ is the strain, and E is the Young's modulus), we obtain (using basic principles [BT]) for linear purely elastic material systems the relationship

$$M(y) = EI\frac{\partial^2 y}{\partial \xi^2}, \tag{5.3}$$

where I represents the moment of inertia of the cross-sectional area. We remark that either E and/or I may be a function of ξ, i.e., $E = E(\xi)$ and/or $I = I(\xi)$.

We note that the derivations underlying the above formulations use the assumptions of linear elasticity and small displacements. Without small displacement assumptions, M could be a nonlinear function of y (for further discussions see [BSW, BT]). Moreover, materials with no internal damping do not exist and thus (5.3) is not an adequate model. We always have internal damping and this must be included in the constitutive relationship (5.2). While there are other forms of internal damping (e.g., spatial hysteresis damping [BaIn] as discussed in Chapter 9), we will assume Kelvin-Voigt damping so that $M = M(y)$ is given by

$$M(y) = EI\frac{\partial^2 y}{\partial \xi^2} + c_D I\frac{\partial^3 y}{\partial \xi^2 \partial t}. \tag{5.4}$$

Combining equations (5.1) and (5.4), we have the beam equation given by

$$\rho\frac{\partial^2 y}{\partial t^2} + \gamma\frac{\partial y}{\partial t} + \frac{\partial^2}{\partial \xi^2}\left(EI\frac{\partial^2 y}{\partial \xi^2} + c_D I\frac{\partial^3 y}{\partial \xi^2 \partial t}\right) = f(t, \xi),$$

with the appropriate boundary conditions

$$\left(EI\frac{\partial^2 y}{\partial \xi^2} + c_D I \frac{\partial^3 y}{\partial \xi^2 \partial t} \right) |_{\xi=l} = 0,$$

$$\left[\frac{\partial}{\partial \xi} \left(EI\frac{\partial^2 y}{\partial \xi^2} + c_D I \frac{\partial^3 y}{\partial \xi^2 \partial t} \right) \right]_{\xi=l} = 0.$$

5.2 The Beam Equation in the Form $\dot{x} = \mathcal{A}x + F$

We wish to write the beam equation in the form $\dot{x} = \mathcal{A}x + F$ for an appropriately defined space X. Let x be represented by

$$x(t, \cdot) = \left(\begin{array}{c} y(t, \cdot) \\ \dot{y}(t, \cdot) \end{array} \right)$$

and take as the state space $X = H_L^2(0, l) \times L_2(0, l)$ where

$$H_L^2(0, l) = \{ \varphi \in H^2(0, l) | \ \varphi(0) = 0, \ \varphi'(0) = 0 \}.$$

Thus for elements in $H_L^2(0, l)$, the function and its derivative both vanish at the left boundary. We rewrite the equation using the abbreviation $\partial = \frac{\partial}{\partial \xi}, \cdot = \frac{\partial}{\partial t}$, and have

$$\ddot{y} = \frac{1}{\rho} \left[-\partial^2 (EI\partial^2 y) - \partial^2 (c_D I \partial^2 \dot{y}) \right] - \frac{\gamma}{\rho} \dot{y} + \frac{1}{\rho} f.$$

We rewrite this in first-order vector form as

$$\frac{d}{dt} \left(\begin{array}{c} y \\ \dot{y} \end{array} \right) = \left(\begin{array}{cc} 0 & I \\ -A_1 & -A_2 \end{array} \right) \left(\begin{array}{c} y \\ \dot{y} \end{array} \right) + \left(\begin{array}{c} 0 \\ \frac{1}{\rho} f \end{array} \right).$$

Let

$$\mathcal{A} = \left(\begin{array}{cc} 0 & I \\ -A_1 & -A_2 \end{array} \right)$$

where we define A_1 and A_2 by

$$A_1 \phi = \frac{1}{\rho} \partial^2 (EI \partial^2 \phi)$$

$$A_2 \phi = \frac{1}{\rho} \partial^2 (c_D I \partial^2 \phi) + \frac{\gamma}{\rho} \phi.$$

Then we have the form $\dot{x} = \mathcal{A}x + F$ in X where

$$\mathcal{D}(\mathcal{A}) = \{ (\varphi, \psi) \in X | \ \psi \in H_L^2(0, l), \ A_1 \varphi + A_2 \psi \in L_2(0, l),$$
$$[EI\partial^2 \varphi + c_D I \partial^2 \psi]_l = 0, \ [\partial(EI\partial^2 \varphi + c_D I \partial^2 \psi)]_l = 0 \}.$$

5.2.1 $\mathcal{A}$ as an Infinitesimal Generator

We claim that $\mathcal{A}$ is an infinitesimal generator of a C_0 semigroup in $X_{\mathcal{E}}$ where $X_{\mathcal{E}}$ is an associated energy space. Specifically, we define $X_{\mathcal{E}}$ as X with the "energy" inner product

$$\langle\langle(\varphi_1,\psi_1),(\varphi_2,\psi_2)\rangle\rangle_{\mathcal{E}} \;=\; \int_0^l EI\partial^2\varphi_1\partial^2\varphi_2 d\xi + \int_0^l \rho\psi_1\psi_2 d\xi$$

$$=\; \int_0^l EI\partial^2\varphi_1\partial^2\varphi_2 d\xi + \langle\rho\psi_1,\psi_2\rangle.$$

If we assume $0 < \rho_1 \leq \rho(\xi) \leq \rho_2 < \infty$, then $\langle\rho\psi_1,\psi_2\rangle$ is equivalent to the norm in $L_2(0,l)$. The norm for $H^2(0,l)$ is defined by

$$|\varphi|_{H^2}^2 = |\varphi|_{L_2}^2 + |\varphi'|_{L_2}^2 + |\varphi''|_{L_2}^2. \tag{5.5}$$

An equivalent norm (this is given as an exercise below) is

$$|\varphi|_{\sim}^2 = |\varphi(a)|^2 + |\varphi'(b)|^2 + |\varphi''|_{L_2}^2, \tag{5.6}$$

where a,b are any values in $[0,l]$. In particular, we have a type of Poincaré Lemma that implies that on H_L^2 we can take the equivalent norm $|\varphi|_{\sim}^2 \equiv |\varphi(0)|^2 + |\varphi'(0)|^2 + |\varphi''|_{L_2}^2 = |\varphi''|_{L_2}^2$.

Lemma 5.1 *In $H_L^2(0,1)$ by choosing $a = b = 0$ in (5.6), we have that $|\varphi''|_{L_2}$ and $|\varphi|_{H^2}$ are equivalent norms. In particular for some constant c we have*

$$|\varphi|_{H^2} \leq c|\varphi''|_{L_2}$$

.

With the corresponding inner product, this can be used to define a Hilbert space $X_{\mathcal{E}}$ which is topologically equivalent to X so that we thus have that if $\mathcal{A}$ generates a C_0 semigroup in $X_{\mathcal{E}}$, the semigroup is also a C_0 semigroup in X. Before proceeding, we state for the sake of completeness an inequality attributed to Poincaré [Adams, p. 159].

Poincaré's (First) Inequality *Suppose there exists a set Ω which is a bounded subset of $\mathbb{R}^n$. Then there exists a constant $c = c(\Omega)$, such that for all $\varphi \in W_0^{k,2}(\Omega) = H_0^k(\Omega)$, one has*

$$|\varphi|_{H^k(\Omega)}^2 \leq c(\Omega) \sum_{|s|=k} \int_\Omega |\partial^s\varphi|^2$$

where $s = (s_1,\ldots,s_n)$.

Homework Exercise

Ex. 9: Use the representation

$$\varphi(\xi) = \varphi(a) + (\xi - a)\varphi'(b) + \int_a^\xi \int_b^\zeta \varphi''(\tau)d\tau d\zeta \qquad (5.7)$$

for $\varphi \in H^2(0, l)$ and a, b arbitrary in $[0, l]$ to argue that $|\cdot|_{H^2}$ and $|\cdot|_\sim$ defined in (5.5) and (5.6), respectively, are equivalent norms on $H^2(0, l)$.

5.2.2   Dissipativeness of $\mathcal{A}$

We argue that $\mathcal{A}$ is dissipative in $X_\mathcal{E}$ under the reasonable physical assumptions that $c_D I(\xi) \geq 0, \gamma(\xi) \geq 0$. Recalling that $M = EI\partial^2\varphi + c_D I\partial^2\psi$, we have on $\mathcal{D}(\mathcal{A})$

$$\langle \mathcal{A}(\varphi, \psi), (\varphi, \psi)\rangle_\mathcal{E} = \int_0^l EI\partial^2\psi\partial^2\varphi + \int_0^l \left[-\partial^2(EI\partial^2\varphi + c_D I\partial^2\psi) - \gamma\psi \right] \psi$$

$$= \int_0^l EI\partial^2\psi\partial^2\varphi - \int_0^l \partial^2(EI\partial^2\varphi + c_D I\partial^2\psi)\psi - \int_0^l \gamma\psi^2)$$

$$= \int_0^l EI\partial^2\psi\partial^2\varphi - \int_0^l ((EI\partial^2\varphi + c_D I\partial^2\psi)\partial^2\psi + \gamma\psi^2)$$
$$- \partial M\psi|_0^l + M\partial\psi|_0^l$$

$$= -\int_0^l c_D I|\partial^2\psi|^2 - \int_0^l \gamma|\psi|^2$$
$$\leq 0.$$

Here $\int_0^l = \int_0^l d\xi$ and we have integrated by parts twice and used the fact that $(\varphi, \psi) \in \mathcal{D}(\mathcal{A})$ so that $\psi(0) = \partial\psi(0) = 0$ and $M(t, l) = \partial M(t, l) = 0$.

5.2.3 $\mathcal{R}(\lambda I - \mathcal{A}) = X$ for some λ

To verify the range statement of Lumer-Phillips, we need to show for some real λ that

$$\lambda(\varphi, \psi) - \mathcal{A}(\varphi, \psi) = (g, h) \qquad (5.8)$$

can be solved for (φ, ψ) for a given $(g, h) \in X$. However, (5.8) reduces to finding, for any $(g, h) \in X = H_L^2 \times L_2$, a solution (φ, ψ) in $\mathcal{D}(\mathcal{A})$ that satisfies

$$-\psi + \lambda\varphi = g \qquad (5.9)$$

$$A_1\varphi + A_2\psi + \lambda\psi = h. \tag{5.10}$$

We can rewrite (5.9) to obtain

$$\psi = \lambda\varphi - g. \tag{5.11}$$

By substituting (5.11) into (5.10), we can reduce the above question to one of solving

$$\lambda^2\varphi + A_1\varphi + \lambda A_2\varphi = h + \lambda g + A_2 g \tag{5.12}$$

for φ, given any (g, h) in $H_L^2(0, l) \times L_2(0, l)$. After solving for φ, we can then obtain ψ using (5.11). If we obtain $(\varphi, \psi) \in \mathcal{D}(\mathcal{A})$, then we have proven the range statement of Lumer-Phillips. To complete the argument we will use the concepts and results for sesquilinear forms to be discussed below (see Sections 6.3, 8.3, and 8.4).

An easy way to solve the above problem is to use the Lax-Milgram theorem; however, we first introduce Gelfand triples. For relevant material, see also [Sh, T, W].

5.3   Gelfand Triples

The usual notation for a Gelfand triple is "$V \hookrightarrow H \hookrightarrow V^*$ with pivot space H." (See the subsection below for more details.) This notation stands for V, H, complex Hilbert spaces, such that $V \subset H$ and V is densely and continuously embedded in H. That is, V is a dense subset of H and

$$|v|_H \leq k|v|_V$$

for all $v \in V$ and some constant k. Therefore, one can identify elements in V with elements in H with an injection operator, the identity i, where i is continuous and $i(V)$ is a dense subset of H.

As explained earlier, we denote by V^* the conjugate dual of V. That is, V^* consists of all conjugate linear continuous functionals on V. (Recall the discussions from Section 1.2.)

For $h \in H$, we define $\varphi(h) \in V^*$ by

$$\varphi(h)(v) = \langle h, v \rangle_H$$

for $v \in V$. We claim that $\varphi : H \to \varphi(H) \subset V^*$ is continuous, linear, one-to-one and onto. Moreover, $\varphi(H)$ is dense in V^* in the V^* topology.

By the Riesz theorem (Theorem 4.2), every $h^* \in H^*$ can be represented by

$$h^*(h) = \langle \hat{h}^*, h \rangle_H \qquad h \in H$$

for some $\hat{h}^* \in H$. Hence, it is readily argued that H^* is isomorphic to H, denoted by $H^* \simeq H$. That is, we may identify H^* with H through $\varphi(H) = \tilde{H} \simeq H^*$ and write $h(v) = \langle h, v \rangle_H$ for $h \in H = H^*$.

This construction is commonly written as $V \hookrightarrow H \simeq H^* \hookrightarrow V^*$ for the pivot space H.

5.3.1 Duality Pairing

With a Gelfand triple, one frequently utilizes the *duality pairing* denoted by $\langle \cdot, \cdot \rangle_{V^*, V}$ given by the extension by continuity of the H inner product from $H \times V$ to $V^* \times V$. That is, for $v^* \in V^*$,

$$v^*(v) = \langle v^*, v \rangle_{V^*, V} = \lim_n \langle h_n, v \rangle_H$$

where $h_n \in H, h_n \to v^*$ in V^*. Note that we have $\langle h, v \rangle_{V^*, V} = \langle h, v \rangle_H$ if $h \in V^*$ also satisfies $h \in H$. Thus we have in effect extended the H inner product from $H \times V$ to $V^* \times V$. (The reader should compare this with the idea of duality product of Definition 4.8.)

5.4 Sesquilinear Forms

Definition 5.1 Let H_1 and H_2 be two complex Hilbert spaces, and let $\sigma : H_1 \times H_2 \to \mathbb{C}$. Then we call σ a *sesquilinear form* if it satisfies

1. σ is linear/conjugate linear (i.e., linear in the first argument, conjugate linear in the second);

2. σ is continuous. More precisely, there exists $\gamma > 0$ such that

$$|\sigma(x, y)| \leq \gamma |x|_{H_1} |y|_{H_2}$$

for $x \in H_1$ and $y \in H_2$.

Definition 5.2 The norm of a sesquilinear form σ is defined by

$$|\sigma| = \sup_{x, y \neq 0} \frac{|\sigma(x, y)|}{|x|_{H_1} |y|_{H_2}}$$

for $x \in H_1$ and $y \in H_2$.

5.4.1   Representations

We consider two different representations:

1. For $y \in H_2$, $x \to \sigma(x,y)$ is continuous and linear on H_1. Therefore, by the Riesz theorem, there exists a unique $z \in H_1$ such
that

$$\sigma(x,y) = \langle x, z \rangle_{H_1}.$$

Thus there exists a mapping $B \in \mathcal{L}(H_2, H_1)$ defined by $By = z$.
Therefore,

$$\sigma(x,y) = \langle x, z \rangle_{H_1} = \langle x, By \rangle_{H_1}.$$

We find that $|\sigma| = |B|_{\mathcal{L}(H_2,H_1)}$. To see this we argue as follows.

For x in H_1 and y in H_2, suppressing notation and using $|\cdot|_1, |\cdot|_2$
for the norms in H_1, H_2, respectively, we have

$$
\begin{aligned}
|\sigma| &= \sup_{x,y\neq 0} \frac{|\sigma(x,y)|}{|x|_1 |y|_2} = \sup_{x,y\neq 0} \frac{|\langle x, By \rangle_{H_1}|}{|x|_1 |y|_2} \\[2mm]
&\leq \sup_{x,y\neq 0} \frac{|x|_1 |By|_1}{|x|_1 |y|_2} = \sup_{y\neq 0} \frac{|By|_1}{|y|_2} \\[2mm]
&= |B|_{\mathcal{L}(H_1,H_2)}.
\end{aligned}
$$

Therefore, we have $|\sigma| \leq |B|_{\mathcal{L}(H_2,H_1)}$. Thus we need only argue
that $|\sigma| \geq |B|_{\mathcal{L}(H_2,H_1)}$. To this end, we have

$$
\begin{aligned}
|\sigma| &= \sup_{x,y\neq 0} \frac{|\sigma(x,y)|}{|x|_1 |y|_2} = \sup_{x,y\neq 0} \frac{|\langle x, By \rangle_{H_1}|}{|x|_1 |y|_2} \\[2mm]
&\geq \sup_{y\neq 0} \frac{|\langle By, By \rangle_{H_1}|}{|By|_1 |y|_2} = \sup_{y\neq 0} \frac{|By|_1}{|y|_2} \\[2mm]
&= |B|_{\mathcal{L}(H_2,H_1)}.
\end{aligned}
$$

Hence we have $|\sigma| = |B|_{\mathcal{L}(H_2,H_1)}$.

2. For $x \in H_1$, $y \to \sigma(x,y)$ is continuous and conjugate linear on
H_2, i.e., $\sigma(x, \cdot) \in H_2^*$. Therefore, by the Riesz theorem, there
exists a unique $z \in H_2$ such that

$$\sigma(x,y) = \langle z, y \rangle_{H_2}.$$

Thus there exists a mapping $A \in \mathcal{L}(H_1, H_2)$ defined by $Ax = z$. Therefore,

$$\sigma(x, y) = \langle z, y \rangle_{H_2} = \overline{\langle Ax, y \rangle_{H_2}},$$

with $|A| = |\sigma|$. Define $\tilde{\sigma}(y, x) = \overline{\sigma(x, y)}$, and interchange the roles of H_1 and H_2 in the above arguments to argue $|A| = |\sigma|$.

We wish to use our Gelfand triple and sesquilinear form formulations to consider elliptic equations of the form $Ax = f$. This gives rise to range statements such as given f, solve for x in $(\lambda - \mathcal{A})x = f$. First, we begin by discussing operators of the form $A : H_1 \rightarrow H_2$ such that A is bounded and continuous. This is very useful in integral equations. However, we will need to modify this formulation to account for the unbounded nature of operators arising in PDE's. We can think of our equation $Ax = f$ in the form $\sigma(x, y) = \langle f, y \rangle_H$ for all y in H where σ is equivalent to A. In other words, $\langle Ax - f, y \rangle_H = 0$ for all y in H. The useful results are given in the famous Lax-Milgram theorems, in bounded and unbounded formulations.

5.5 Lax-Milgram—bounded form

Theorem 5.1 *(Lax-Milgram theorem-bounded form) Suppose $\sigma : H \times H \rightarrow \mathbb{C}$ is continuous and linear/conjugate linear, i.e., it is a sesquilinear form with*

$$|\sigma(x, y)| \leq \gamma |x|_H |y|_H$$

for all x and y in H, and σ is strictly positive, i.e.,

$$|\sigma(x, x)| \geq \delta |x|^2$$

for all x in H where γ and δ are positive constants. Then there exists $A : H \rightarrow H, A \in \mathcal{L}(H)$ defined by

$$\sigma(x, y) = \langle Ax, y \rangle$$

with $|A| = |\sigma| \leq \gamma$ for all x and y in H. Moreover, $A^{-1} \in \mathcal{L}(H)$ exists and

$$\sigma(A^{-1}x, y) = \langle x, y \rangle$$

with $|A^{-1}| \leq \frac{1}{\delta}$ for all x and y in H.

Proof
By the Riesz theorem, we know there exists $A \in \mathcal{L}(H)$ such that $|A| = |\sigma| \leq \gamma$ where $\sigma(x, y) = \langle Ax, y \rangle_H$ for all x and y in H.

We claim that A is one-to-one. For this we need to show that $Ax = 0$ implies $x = 0$. Suppose $Ax = 0$. Then, by the definition of A and the assumptions given, we have

$$
\begin{aligned}
0 &= |\langle Ax, x \rangle| = |\sigma(x, x)| \\
&\geq \delta |x|^2,
\end{aligned}
$$

which implies $|x| = 0$, and hence $x = 0$. In other words, A is one-to-one. Since A is one-to-one, we know that A^{-1} exists on $\mathcal{R}(A) \subset H$.

We next claim that A^{-1} is bounded on $\mathcal{R}(A)$. By the proposition that σ was strictly positive and the definition of A, we have

$$
\begin{aligned}
\delta |x|^2 &\leq |\sigma(x, x)| = |\langle Ax, x \rangle| \\
&\leq |Ax||x|.
\end{aligned}
$$

Therefore, $|Ax| \geq \delta |x|$ for all x in H. In particular, for $x = A^{-1}y$, we have

$$ |y| \geq \delta |A^{-1}y|. $$

Therefore,

$$ |A^{-1}y| \leq \frac{1}{\delta}|y| $$

for all y in $\mathcal{R}(A)$. In other words, A^{-1} is bounded. Finally all we have to show is that $\mathcal{R}(A) = H$. By assumption, we know that A is a continuous operator and we have argued that A^{-1} is bounded on $\mathcal{R}(A)$. Therefore, $\mathcal{R}(A)$ is a closed subspace of H. Now suppose $\mathcal{R}(A) \neq H$. This implies, by orthogonal decomposition results for closed subspaces in Hilbert spaces, there exists $z \neq 0$ such that $z \perp \mathcal{R}(A)$. In other words,

$$ \langle Ax, z \rangle = 0 $$

for all x in H. In particular, choosing $x = z$, we have

$$ 0 = \langle Az, z \rangle = \sigma(z, z) \geq \delta |z|^2. $$

This implies $z = 0$; therefore, we have a contradiction. So, $\mathcal{R}(A) = H$.

5.5.1   Discussion of $Ax = f$ with A bounded

The bounded form of Lax-Milgram, while not very useful in studying partial differential equations, is very useful in linear integral equations.

The Fredholm integral equations with kernel $k \in L_2([a,b] \times [a,b])$ are given by

$$\int_a^b k(\zeta, \xi)\varphi(\xi)d\xi = f(\zeta) \qquad \zeta \in [a,b] \qquad \text{(first kind)}$$

$$\varphi(\zeta) - \int_a^b k(\zeta, \xi)\varphi(\xi)d\xi = f(\zeta) \qquad \zeta \in [a,b] \qquad \text{(second kind)}.$$

These can be written as operator equations in $H = L^2(a,b)$,

$$A\varphi = f$$
$$\varphi - A\varphi = f$$

to be solved for φ, given f, where A is a bounded linear operator (discussed below) in H. It also has applications in scattering theory (acoustic and electromagnetic radiation), far field radiation, and single and double layer potential theory. For relevant material, see also [K, CK1, CK2].

As we have noted, the bounded form of Lax-Milgram is adequate if one wants to solve $Ax = f$ in H where A is bounded. Consider $k \in L_2(\Omega)$ with $\Omega = [0,1] \times [0,1]$ with $H = L_2(0,1)$. Then define $A : H \to H$ by

$$(A\varphi)(\zeta) = \int_0^1 k(\zeta, \xi)\varphi(\xi)d\xi.$$

We can see that this operator is bounded, i.e.,

$$\int_0^1 \left(\int_0^1 k(\zeta, \xi)\varphi(\xi)d\xi \right)^2 d\zeta \le \gamma \int_0^1 |\varphi(\xi)|^2 d\xi.$$

Therefore, there is a sesquilinear form σ such that A corresponds with σ as in the Lax-Milgram above with

$$\sigma(\varphi, \psi) = \int_0^1 \left(\int_0^1 k(\zeta, \xi)\varphi(\xi)d\xi \right) \psi(\zeta)d\zeta$$

$$= \langle A\varphi, \psi \rangle_{L_2(0,1)}.$$

We can show that all the requirements of the Lax-Milgram theorem (bounded form) are met with this operator A; therefore, the Lax-Milgram theorem (bounded form) is applicable.

5.5.2 Example–The Steady State Heat Equation in $H_0^1(\Omega)$

Let $\Omega = [0,1] \times [0,1]$. Then the heat equation is given by

$$\frac{\partial u}{\partial t} = \nabla \cdot (D\nabla u) + f$$

where $\nabla = \frac{\partial}{\partial \xi_1}\hat{i} + \frac{\partial}{\partial \xi_2}\hat{j}$ with $(\xi_1, \xi_2) \in \Omega$. We assume boundary conditions that require $u \in H_0^1(\Omega)$. Then the steady state equation is given by

$$-\nabla \cdot (D\nabla u) = f \tag{5.13}$$

or

$$-\left(\frac{\partial}{\partial \xi_1}(D\frac{\partial u}{\partial \xi_1}) + \frac{\partial}{\partial \xi_2}(D\frac{\partial u}{\partial \xi_2}) \right) = f(\xi_1, \xi_2).$$

In the weak or variational form, we have (by integration by parts and use of the boundary conditions)

$$\langle D\nabla u, \nabla \varphi \rangle_{L_2} = \langle f, \varphi \rangle_{L_2} \tag{5.14}$$

for a test function $\varphi \in H_0^1(\Omega)$. We define σ on $H_0^1 \times H_0^1$ by $\sigma(\psi, \varphi) = \langle D\nabla \psi, \nabla \varphi \rangle_{L_2}$ with $|D|_\infty \leq \gamma$, i.e., D is bounded. Then σ is not continuous on $H = L_2(\Omega)$. On the other hand, for $\varphi, \psi \in H_0^1$

$$|\sigma(\psi, \varphi)| \leq \gamma |\nabla \psi|_{L_2} |\nabla \varphi|_{L_2}$$

$$\leq \gamma |\psi|_{H_0^1} |\varphi|_{H_0^1}$$

implies σ is continuous on $V = H_0^1(\Omega)$.

Moreover, we do have some type of positivity. If $|D(\xi_1, \xi_2)| \geq \delta$, then

$$|\sigma(\varphi, \varphi)| = |\langle D\nabla \varphi, \nabla \varphi \rangle_{L_2}|$$

$$\geq \delta |\nabla \varphi|_{L_2}^2 \geq \tilde{\delta} |\varphi|_{H_0^1(\Omega)}.$$

In other words, in the $V = H_0^1(\Omega)$ norm, we would have both continuity and strict positivity of the sesquilinear form. But we do not have continuity and strict positivity in the $H = L_2(\Omega)$ sense. Thus if we choose our H space to be $H_0^1(\Omega)$, we would be guaranteed a unique solution of $\langle Au - f, \varphi \rangle_{H_0^1} = 0$. That is, $u = A^{-1}f$ in the H_0^1 sense (and of course $u \in H_0^1$ since A^{-1} is bounded from H_0^1 to H_0^1). This also requires the input function f to be in $H_0^1(\Omega)$ which is a strong requirement for f. Therefore, it would not be useful in some cases to choose our H space to be $H^1_0(\Omega)$ in the Lax-Milgram theorem. Instead, we need an extension of Lax-Milgram to treat the heat equation in a more reasonable manner.

5.6   Lax-Milgram—unbounded form

Let $V \hookrightarrow H \hookrightarrow V^*$ be a Gelfand triple.

Definition 5.3 A sesquilinear form $\sigma : V \times V \to \mathbb{C}$ is said to be V-*continuous* if

$$|\sigma(\varphi, \psi)| \leq \gamma |\varphi|_V |\psi|_V$$

for all φ and ψ in V.

Consequences of V-continuity

For a fixed φ in V, we consider the mapping $\psi \to \sigma(\varphi, \psi)$ for a V-continuous σ. This mapping is continuous by the definition of continuity above, and it is a conjugate linear mapping into $\mathbb{C}$. Therefore, $\psi \to \sigma(\varphi, \psi)$ is in V^*. This implies there exists an operator $\mathcal{A} \in \mathcal{L}(V, V^*)$ such that $\sigma(\varphi, \psi) = \langle \mathcal{A}\varphi, \psi \rangle_{V^*, V}$.

Conversely, if $\mathcal{A} \in \mathcal{L}(V, V^*)$, we can define $\sigma : V \times V \to \mathbb{C}$ by $\sigma(\varphi, \psi) = (\mathcal{A}\varphi)\psi$ so that σ is V-continuous and linear/conjugate linear.

In other words, if we have a V-continuous sesquilinear form σ, there is a one-to-one correspondence between σ and $\mathcal{A} \in \mathcal{L}(V, V^*)$. While the operator is not bounded on V, it will be useful in considering the equation $\mathcal{A}u = f$ in V^*. (Note the weak requirements on f for example in the case $V = H^1_0$ and $V^* = H^{-1}$.)

Definition 5.4 A sesquilinear form σ is V-*coercive* if there exists a constant $\delta > 0$ such that

$$|\sigma(\varphi, \varphi)| \geq \delta |\varphi|^2_V$$

for $\varphi \in V$.

Theorem 5.2 (*Lax-Milgram theorem-unbounded form*) *Let* $V \hookrightarrow H \hookrightarrow V^*$ *be a Gelfand triple. Let* $\sigma : V \times V \to \mathbb{C}$ *be a V-continuous, V-coercive sesquilinear form. Then* $\mathcal{A} : V \to V^*$ *given by*

$$\sigma(\varphi, \psi) = \langle \mathcal{A}\varphi, \psi \rangle_{V^*, V}$$

is a linear (topological) isomorphism between V and V^*. Moreover, $\mathcal{A}^{-1}$ is continuous from V^* to V with

$$|\mathcal{A}^{-1}|_{\mathcal{L}(V^*,V)} \leq \frac{1}{\delta}.$$

Proof

Let $R : V^* \to V$ be a Riesz isomorphism, so that for $v^* \in V^*$ we have

$$v^*(v) = \langle v^*, v \rangle_{V^*,V} = \langle Rv^*, v \rangle_V. \tag{5.15}$$

Then $R^{-1} : V \to V^*$ is continuous.

Let $\sigma : V \times V \to \mathbb{C}$ be a V-continuous, V-coercive sesquilinear form. Then $\psi \to \sigma(\varphi, \psi)$ in V^* implies there exists a z in V such that $\sigma(\varphi, \psi) = \langle z, \psi \rangle_V$. From the bounded version of Lax-Milgram, this implies there exists $A \in \mathcal{L}(V)$ such that

$$\sigma(\varphi, \psi) = \langle A\varphi, \psi \rangle_V \tag{5.16}$$

for all φ and ψ in V, with A one-to-one, A onto, $|A|_{\mathcal{L}(V)} \leq \gamma$ and $|A^{-1}|_{\mathcal{L}(V)} \leq \frac{1}{\delta}$. In other words, A is an isomorphism $V \to V$.

We have that $R^{-1} : V \to V^*$ is also an isomorphism; therefore, $R^{-1}A : V \to V^*$ is an isomorphism. The claim is that $\mathcal{A} = R^{-1}A$. By (5.15) we have for all $\varphi, \psi \in V$

$$\langle R^{-1}A\varphi, \psi \rangle_{V^*,V} = \langle A\varphi, \psi \rangle_V.$$

However, by (5.16), this implies

$$\langle R^{-1}A\varphi, \psi \rangle_{V^*,V} = \sigma(\varphi, \psi).$$

Therefore, by definition of $\mathcal{A} : V \to V^*$, $\mathcal{A}$ must be given by $\mathcal{A} = R^{-1}A$. So, we have

$$\langle \mathcal{A}\varphi, \psi \rangle_{V^*,V} = \sigma(\varphi, \psi) \leq \gamma |\varphi|_V |\psi|_V \tag{5.17}$$

and

$$\langle \mathcal{A}\varphi, \varphi \rangle_{V^*,V} \geq \delta |\varphi|_V^2. \tag{5.18}$$

Note that $\sup_{|\psi|=1} |\langle \mathcal{A}\varphi, \psi \rangle_{V^*,V}| = |\mathcal{A}\varphi|_{V^*}$. Letting $\varphi = \mathcal{A}^{-1}\psi$ in (5.18) and combining this with (5.17) and (5.18), we have

$$\delta |\varphi|_V \leq |\mathcal{A}\varphi|_{V^*} \leq \gamma |\varphi|_V. \tag{5.19}$$

Therefore, we have

$$|\mathcal{A}|_{\mathcal{L}(V,V^*)} \leq \gamma$$

and

$$|\mathcal{A}^{-1}|_{\mathcal{L}(V^*,V)} \leq \frac{1}{\delta}.$$

Implications of Lax-Milgram (unbounded form)

For $\mathcal{A}$ as in the Lax-Milgram theorem 5.2 and $f \in V^*$, we consider $\mathcal{A}u = f$ in V^*. Then Lax-Milgram implies there exists a unique solution $u = \mathcal{A}^{-1}f$ in V that depends continuously on f. More precisely,

$$|u|_V = |\mathcal{A}^{-1}f|_V \leq \frac{1}{\delta}|f|_{V^*}.$$

We revisit the steady-state heat equation (5.13) in the example above with $D \in L_\infty(\Omega)$ and $V = H_0^1(\Omega)$. This, of course, allows discontinuous coefficients. Then for any f in $V^* = H^{-1}(\Omega)$ there exists a unique $u \in V$ satisfying $\mathcal{A}u = f$. That is, $\langle \mathcal{A}u - f, \varphi \rangle_{V^*,V} = \sigma(u,\varphi) - \langle f, \varphi \rangle_{V^*,V} = \langle D\nabla u, \nabla \varphi \rangle_{L_2} - \langle f, \varphi \rangle_{V^*,V} = 0$ for all $\varphi \in V$. Thus we say $u \in V$ satisfies $\mathcal{A}u = f$ in the sense of V^*. Hence (5.13) holds in the V^* sense which is precisely (5.14) with $\langle f, \varphi \rangle$ in the more general form $\langle f, \varphi \rangle_{V^*,V}$ which permits f in the weaker sense $f \in V^* = H^{-1}$. This is also sometimes referred to as u satisfying (5.13) in the *sense of distributions*.

5.6.1 The Concept of D_A

We assume that we have a Gelfand triple $V \hookrightarrow H \hookrightarrow V^*$ and a sesquilinear form $\sigma : V \times V \to \mathbb{C}$. As usual, $f(\psi) = \langle f, \psi \rangle_H$ defines, for $f \in H$, an element $f \in V^*$. Considering $(\mathcal{A}\varphi)(\psi) = \sigma(\varphi, \psi)$ for φ and ψ in V, we define

$$D_A = \{\varphi \in V | \mathcal{A}\varphi \in H\}.$$

That is, D_A is the set of $\varphi \in V$ such that $\mathcal{A}\varphi \in V^*$ has the representation $(\mathcal{A}\varphi)(\psi) = \langle \tilde{\varphi}, \psi \rangle_H$, for ψ in V and for some $\tilde{\varphi}$ in H.

We denote this element $\tilde{\varphi}$ by $-A\varphi = \tilde{\varphi}$, i.e., A is linear from $D_A \subset V$ into H and given by

$$\sigma(\varphi, \psi) = \langle -A\varphi, \psi \rangle_H$$

for ψ in V and φ in D_A.

We note that the above can be interpreted as: $\varphi \in D_A \subset V$ if and only if $\varphi \in V$ and $\mathcal{A}\varphi \in H^* \cong H$ so that $(\mathcal{A}\varphi)(\psi) = \langle \tilde{\varphi}, \psi \rangle_H$, for all ψ in V, and for some $\tilde{\varphi} \in H$. Alternatively, we may write

$$D_A = \{\varphi \in V | |(\mathcal{A}\varphi)(\psi)| = |\sigma(\varphi, \psi)| \leq k|\psi|_H, \psi \in V\}$$

for some k. Moreover, we could write

$$D_A = \{\varphi \in V | \psi \to \sigma(\varphi, \psi) \text{ is in } H^*, \text{ i.e., continuous on } H\}.$$

Note that we also have

$$\sigma(\varphi, \psi) = (A\varphi)(\psi) = \langle A\varphi, \psi \rangle_{V^*,V}$$

for all φ and ψ in V. If we restrict $\varphi \in D_A$ then this also equals $\langle -A\varphi, \psi \rangle_H$.

Theorem 5.3 *If σ is a V-continuous V-coercive sesquilinear form on V, then D_A is dense in V and, hence, dense in H.*

Proof Define $\tilde{\sigma}(\varphi, \psi) = \overline{\sigma(\psi, \varphi)}$ (called the "adjoint" sesquilinear form).

If σ is V-coercive and V-continuous, then $\tilde{\sigma}$ is also V-coercive and V-continuous. In other words, there exists an operator $\tilde{A} : V \to V^*$ such that $\tilde{A} \in \mathcal{L}(V, V^*)$ with

$$\tilde{\sigma}(\varphi, \psi) = \langle \tilde{A}\varphi, \psi \rangle_{V^*,V} = (\tilde{A}\varphi)(\psi)$$

for all φ and ψ in V. Hence, $\mathcal{R}(\tilde{A}) = V^*$ by the Lax-Milgram theorem.

Next we need to show that D_A is dense in V. It suffices to show that if $f \in V^*$ and $f(v) = 0$ for all $v \in D_A$, then $f \equiv 0$. Let $f \in V^*$ be such that $f(v) = 0$ for all $v \in D_A$. Since $f \in V^*$ and $\mathcal{R}(\tilde{A}) = V^*$, then there exists $\varphi \in V$ such that $f = \tilde{A}\varphi$.
For $v \in D_A$, we have

$$
\begin{aligned}
(-Av)(\varphi) &= (Av)(\varphi) = \sigma(v, \varphi) \\
&= \overline{\tilde{\sigma}(\varphi, v)} = \overline{\langle \tilde{A}\varphi, v \rangle_{V^*,V}} \\
&= \overline{\langle f, v \rangle_{V^*,V}} = \overline{f(v)} \\
&= 0.
\end{aligned}
$$

Therefore, $(Av)(\varphi) = 0$ for all $v \in D_A$. But, we know $\mathcal{R}(A|_{D_A}) = H \cong H^*$. Then, for every $h \in H^*$, $h(\varphi) = 0$. However, as H^* is dense in V^*, then this implies $\varphi = 0$. Moreover, $\tilde{A}\varphi = f$ implies $f = 0$.

Thus we find that any V-continuous V-coercive sesquilinear form on V gives rise to a densely defined operator A on $\mathcal{D}(A) = D_A$ with

$$
\begin{aligned}
\sigma(\varphi, \psi) &= \langle A\varphi, \psi \rangle_{V^*,V} \quad \varphi, \psi \in V \\
&= \langle -A\varphi, \psi \rangle_H \quad \varphi \in D_A, \psi \in V.
\end{aligned}
$$

5.6.2   V-elliptic

Definition 5.5 A sesquilinear form σ on V is *V-elliptic* if there exists a constant $\delta > 0$ such that

$$\operatorname{Re} \sigma(\varphi, \varphi) \geq \delta |\varphi|_V^2 \quad \varphi \in V.$$

Discussion of V-elliptic

We note that σ is V-elliptic implies σ is V-coercive. Of course, if we are working in real spaces, $\operatorname{Re} \sigma = \sigma$ and V-elliptic is equivalent to V-coercive. We also remark that the terminology among various authors is not standard. Some authors (e.g., Wloka in [W]) use our definition for V-coercive as the definition of V-elliptic and then use

$$|\sigma(\varphi, \varphi) + k|\varphi|_H^2| \geq \delta |\varphi|_V^2 \quad \varphi \in V, \delta > 0$$

as the definition of V-coercive.

Some of the terminology and usage of sesquilinear forms derives directly from that for PDE's of the form

$$\frac{\partial y}{\partial t} = \sum_{i,j} \frac{\partial}{\partial \xi_i} \left(a_{ij} \frac{\partial y}{\partial \xi_j} \right) + \sum_j b_j \frac{\partial y}{\partial \xi_j}, \quad t > 0, \xi \in G \subset \mathbb{R}^n.$$

Definition 5.6 In classical PDE's, an "operator" $\{a_{ij}\}$ is said to be *strongly elliptic* on G if there exists $\delta > 0$ such that for $\xi \in G$

$$\operatorname{Re} \sum_{i,j} a_{ij}(\xi) q_i \bar{q}_j \geq \delta \sum_i |q_i|^2$$

for all $q \in \mathbb{C}^n$.

An associated sesquilinear form on $V = H^1(G)$ can be defined by

$$\sigma(\varphi, \psi) = \int_G \left[\sum_{i,j} a_{ij}(\xi) \frac{\partial \varphi}{\partial \xi_i} \frac{\partial \bar{\psi}}{\partial \xi_j} + \sum_k b_k \frac{\partial \varphi}{\partial \xi_k} \bar{\psi} \right] d\xi.$$

Discussion of Strongly Elliptic

It is a standard result [Sh, p.65] that $\{a_{ij}\}$ strongly elliptic implies there exists $\delta > 0$ such that for λ sufficiently large

$$\operatorname{Re} \sigma(\varphi, \varphi) + \lambda |\varphi|_H^2 \geq \delta |\varphi|_V^2, \quad \varphi \in V.$$

Hence if $\{a_{ij}\}$ is strongly elliptic, then for some λ_0 sufficiently large,

$$\tilde{\sigma}(\varphi, \psi) = \sigma(\varphi, \psi) + \lambda_0 \langle \varphi, \psi \rangle_H$$

is V-elliptic and hence V-coercive.

A most useful result (Theorems 6.1 and 6.2) is that continuous V-elliptic forms give rise to operators that are infinitesimal generators of C_0 (actually analytic) semigroups.

5.7 Summary Remarks and Motivation

We used the beam equation of Sections 5.1–5.2 and in particular the need to solve range equations such as (5.8)–(5.12) in semigroup generation theorems (Lumer-Phillips) to motivate development of the Lax-Milgram lemmas in the context of sesquilinear forms. We will use these results in the study of general second-order systems in Chapter 8. Before turning to these results we present in the next chapter results on stronger semigroup generation theorems (also applicable to the beam equation). For the motivating beam equation we shall see the strength of damping in the equation will determine which of the generation theorems (and hence regularity of the generated semigroup) can be used to guarantee results for the beam equation (specifically in Sections 8.2–8.4—see also the discussions of damping operators in Section 9.2).

6 Analytic Semigroups

To this point we have discussed strongly continuous (C_0) semigroups and associated generation theorems. For some examples we have already discussed one actually obtains a stronger regularity (analyticity) than continuity and this will prove most useful in our discussion of infinite dimensional control theory later (Chapter 16). We digress slightly to discuss results that will enable one to guarantee this stronger regularity in semigroups.

Definition 6.1 A semigroup $T(t)$ on H is called an *analytic semigroup* if $t \to T(t)\varphi$ is analytic for each φ in H.

Theorem 6.1 *Let V, H be complex Hilbert spaces with $V \hookrightarrow H \hookrightarrow V^*$ and suppose that $\sigma : V \times V \to \mathbb{C}$ is V-continuous and V-elliptic; i.e.*

$$|\sigma(\varphi, \psi)| \leq \gamma |\varphi|_V |\psi|_V \quad \varphi, \psi \in V,$$

and

$$\text{Re } \sigma(\varphi, \varphi) \geq \delta |\varphi|_V^2 \quad \delta > 0, \varphi \in V.$$

Define $A : \mathcal{D}(A) \subset V \to H$ by

$$
\begin{aligned}
\mathcal{D}(A) \quad = \quad & \{\varphi \in V | \text{ there exists } K_\varphi > 0 \text{ such that} \\
& |\sigma(\varphi, \psi)| \leq K_\varphi |\psi|_H, \psi \in V\}
\end{aligned}
$$

and

$$\sigma(\varphi, \psi) = \langle -A\varphi, \psi \rangle_H, \quad \varphi \in \mathcal{D}(A), \psi \in V.$$

Then $\mathcal{D}(A)$ is dense in H and A is the infinitesimal generator of a contraction semigroup in H that actually is an analytic semigroup.

The proof of this theorem can be found in [Sh, Thm 6.A, p. 99], while the next result is an immediate consequence of this theorem. Note that by the remarks in Section 5.6.1, the $\mathcal{D}(A)$ in this theorem is the same as D_A defined there even though each definition and notation is found in the literature.

Theorem 6.2 *Suppose all the assumptions of Theorem 6.1 hold except that the V-ellipticity condition for σ is replaced by*

$$\text{Re } \sigma(\varphi, \varphi) + \lambda_0 |\varphi|_H^2 \geq \delta |\varphi|_V^2$$

for some $\lambda_0, \delta > 0, \varphi \in V$. Then defining A as in Theorem 6.1, we have that A is densely defined and is the infinitesimal generator of an analytic semigroup in H.

6.1   Example 1: The Heat Equation (again)

The system is given by

$$\frac{\partial y}{\partial t} = \frac{\partial}{\partial \xi}\left(D(\xi)\frac{\partial y}{\partial \xi}\right)$$

$$y(t,0) = 0, \quad \frac{\partial y}{\partial \xi}(t,l) = 0$$

$$y(0,\xi) = \Phi(\xi).$$

We choose the state space $X = H = L_2(0,l)$ as before. To obtain the weak variational form and the space V, we work backwards by multiplying the equation by a "test" function φ and integrating. This yields

$$\int_0^l \dot{y}\varphi = \int_0^l (Dy')'\varphi$$

$$= \int_0^l -Dy'\varphi' + Dy'\varphi|_0^l$$

Therefore we have

$$\langle \dot{y}(t), \varphi \rangle + \langle Dy'(t), \varphi' \rangle - Dy'(t)\varphi|_0^l = 0 \tag{6.1}$$

where $\langle \cdot, \cdot \rangle$ is the usual inner product in L_2. However, (6.1) is equivalent to

$$\langle \dot{y}(t), \varphi \rangle + \langle Dy'(t), \varphi' \rangle = 0$$

if $\varphi \in H_L^1(0,l) = \{\varphi \in H^1(0,l)|\varphi(0) = 0\}$ and $Dy'(t,l) = 0$ (the right boundary condition at $\xi = l$) .

Defining $V = H_L^1(0,l)$ and σ on $V \times V$ by

$$\sigma(\varphi, \psi) = \langle D\varphi', \psi' \rangle,$$

we may write the equation in *weak form* as: find $y(t) \in V$ satisfying

$$\langle \dot{y}(t), \varphi \rangle + \sigma(y(t), \varphi) = 0$$

for all $\varphi \in V$. This equation is equivalent to the original system whenever $y(t) \in V \bigcap H^2(0,l)$ by using the reverse of the above arguments.

That is, a weak form solution will also be a solution of the original (classical or strong form) equation if it has sufficient smoothness and satisfies the flux boundary conditions at $\xi = l$.

To complete verification of this equivalence, we next consider the flux boundary condition of the original problem. Suppose $y(t)$ is a weak solution, i.e.,

$$\langle \dot{y}(t), \varphi \rangle + \sigma(y(t), \varphi) = 0 \quad \text{for all } \varphi \in V$$

$$y(0) = \Phi(\xi)$$

and y in $H^2(0, l)$. Then

$$\langle \dot{y}(t), \varphi \rangle + \int_0^l Dy'\varphi' = 0.$$

Integrating by parts, we find that the above equation is equivalent to

$$\int_0^l (\dot{y} - (Dy')')\varphi + Dy'(t, l)\varphi(l) = 0 \qquad (6.2)$$

for all $\varphi \in H_L^1$. However, $H_0^1 \subset H_L^1$; therefore,

$$\int_0^l (\dot{y} - (Dy')')\varphi = 0 \qquad (6.3)$$

for all $\varphi \in H_0^1$. Since H_0^1 is dense in $L_2(0, l)$, this implies $\dot{y} - (Dy')' = 0$. However, if we choose $\varphi \in H_L^1$ such that $\varphi(l) \neq 0$, then (6.2) implies $Dy'(t, l) = 0$, i.e., the flux boundary condition is satisfied for weak solutions with additional smoothness. A relationship between this concept of weak solution and semigroup solutions will be given in a subsequent section (see Chapter 7).

The above discussions motivate the following formulation using sesquilinear forms. Define the V-inner product as

$$\langle \varphi, \psi \rangle_V = \int_0^l \varphi'\psi',$$

and recall $H = X = L_2(0, l)$; then we readily see $V \hookrightarrow H \hookrightarrow V^*$. Note that the V norm is equivalent to the usual H^1 norm on $H_L^1(0, l)$ by the Poincaré type result of Lemma 5.1. Furthermore, we have

$$|\sigma(\varphi, \psi)| = |\langle D\varphi', \psi' \rangle|$$

$$\leq |D|_\infty |\varphi'|_{L^2} |\psi'|_{L^2}$$

$$= |D|_\infty |\varphi|_V |\psi|_V.$$

Also,

$$\mathrm{Re}\ \sigma(\varphi, \varphi) = \ \mathrm{Re}\ \langle D\varphi', \varphi'\rangle \geq \delta|\varphi'|_{L_2}^2 = \delta|\varphi|_V^2$$

so that σ is V-continuous and V-elliptic.

We can define $\mathcal{A} : V \to V^*$ by

$$\langle \mathcal{A}\varphi, \psi\rangle_{V^*,V} = \sigma(\varphi, \psi) = \langle D\varphi', \psi'\rangle.$$

Note that $\mathcal{A}\varphi \in H \hookrightarrow V^*$ if and only if $\langle D\varphi', \psi'\rangle = \langle w, \psi\rangle$ for some $w \in H$ and for all $\psi \in V$. However, integrating by parts we have

$$\int_0^l D\varphi'\psi' = -\int_0^l (D\varphi')'\psi + D\varphi'\psi|_0^l$$

$$= \langle -(D\varphi')', \psi\rangle_H + D(l)\varphi'(l)\psi(l)$$

$$= \langle -(D\varphi')', \psi\rangle_H$$

if $\varphi'(l) = 0$ and $(D\varphi')' \in L_2(0, l)$. Thus we may define

$$\mathcal{A}\varphi = (D\varphi')'$$

on

$$\mathcal{D}(A) = \{\varphi \in H_L^1(0, l)|(D\varphi')' \in L_2(0, l), \varphi'(l) = 0\}$$

and obtain $\mathcal{A}\varphi = -A\varphi \in H$ exactly whenever $\varphi \in \mathcal{D}(A)$.

The above results hence guarantee that A generates a C_0-semigroup (actually an analytic semigroup–Theorem 6.1) $T(t)$ on $H = X = L_2(0, l)$.

6.2 Example 2: The Transport Equation (again)

We consider again the transport equation given by

$$\begin{aligned}
\frac{\partial y}{\partial t} + \frac{\partial}{\partial \xi}(\nu y) &= \frac{\partial}{\partial \xi}\left(D\frac{\partial y}{\partial \xi}\right) - \mu y \\
y(t, 0) &= 0 \\
(D\frac{\partial y}{\partial \xi} - \nu y)|_{\xi=l} &= 0 \\
y(0, \xi) &= \Phi(\xi).
\end{aligned} \tag{6.4}$$

We can rewrite the transport equation as

$$y_t = (Dy' - \nu y)' - \mu y.$$

Multiplying by a test function and integrating from 0 to l, we have

$$\langle y_t, \varphi \rangle = \int_0^l ((Dy' - vy)'\varphi - \mu y\varphi)\, d\xi$$

$$= -\langle Dy' - vy, \varphi' \rangle + (Dy' - vy)\varphi|_0^l - \langle \mu y, \varphi \rangle.$$

If we choose $H = X = L_2(0,l)$ and $V = H_L^1(0,l)$ as in Example 1 above, with the same V inner product, we have

$$\langle y_t, \varphi \rangle = -\langle Dy' - vy, \varphi' \rangle - \langle \mu y, \varphi \rangle.$$

As before, we have $V \hookrightarrow H \hookrightarrow V^*$. Then we can define the sesquilinear form $\sigma : V \times V \to C$ by

$$\sigma(\varphi, \psi) = \langle D\varphi' - v\varphi, \psi' \rangle + \langle \mu\varphi, \psi \rangle.$$

Therefore, we have the equation

$$\langle \dot{y}, \varphi \rangle + \sigma(y, \varphi) = 0. \tag{6.5}$$

We digress briefly to discuss the various other possibilities for boundary conditions and the effects on the choice of V in this transport equation example. If we had a no flux boundary condition at $\xi = 0$ and $y(t,l) = 0$, we would choose $V = H_R^1(0,l) \equiv \{\varphi \in H^1(0,l) \mid \varphi(l) = 0\}$. On the other hand, if we had essential boundary conditions at both boundaries, i.e., $y = 0$ at $\xi = 0$ and $\xi = l$, we would need to choose $V = H_0^1(0,l)$. A third possibility is if we had the no flux boundary conditions at both boundaries, $\xi = 0$ and $\xi = l$. In that case, as both boundary conditions were natural, we would choose $V = H^1(0,l)$.

Returning to equations (6.4) and (6.5), we can establish V-continuity of σ by arguing

$$|\sigma(\varphi,\psi)| \le |D|_\infty|\varphi'|_H|\psi'|_H + |v|_\infty|\varphi|_H|\psi'|_H + |\mu|_\infty|\varphi|_H|\psi|_H$$

$$\le |D|_\infty|\varphi|_V|\psi|_V + |v|_\infty k|\varphi|_V|\psi|_V + |\mu|_\infty k^2|\varphi|_V|\psi|_V$$

$$= (|D|_\infty + k|v|_\infty + k^2|\mu|_\infty)|\varphi|_V|\psi|_V.$$

As σ is V-continuous, we have

$$\sigma(\varphi,\psi) = \langle \mathcal{A}\varphi, \psi \rangle_{V^*,V} \quad \varphi \in V$$
$$= \langle -A\varphi, \psi \rangle_H \quad \varphi \in \mathcal{D}(A),$$

where $\mathcal{D}(A)$ is defined by

$$\mathcal{D}(A) = \{\varphi \in H^2(0, l) | \varphi(0) = 0, (D\varphi' - \nu\varphi) \in H^1(0, l), (D\varphi' - \nu\varphi)(l) = 0\}.$$

Note that V carries the essential boundary conditions, while the natural boundary conditions are found in $\mathcal{D}(A)$.

To show that σ is V-elliptic (or actually a shift or translation of σ is V-elliptic), we assume $D \geq c_1 > 0$ and $\langle \mu\varphi, \varphi \rangle \geq -|\mu|_\infty |\varphi|_H^2$; then we have

$$\text{Re } \sigma(\varphi, \varphi) \geq c_1 |\varphi|_V^2 - \frac{|\nu|_\infty^2}{4\epsilon} |\varphi|_H^2 - \epsilon |\varphi|_V^2 - |\mu|_\infty |\varphi|_H^2$$

$$= (c_1 - \epsilon) |\varphi|_V^2 - (\frac{|\nu|_\infty^2}{4\epsilon} + |\mu|_\infty) |\varphi|_H^2.$$

Hence, setting $\epsilon = \frac{c_1}{2}$, we have

$$\text{Re } \sigma(\varphi, \varphi) \geq \frac{c_1}{2} |\varphi|_V^2 - \lambda_0 |\varphi|_H^2$$

for $\lambda_0 = \frac{|\nu|_\infty^2}{2c_1} + |\mu|_\infty$. Thus we see that $\tilde{\sigma}$ given by

$$\begin{aligned}
\tilde{\sigma}(\varphi, \psi) &= \sigma(\varphi, \psi) + \lambda_0 \langle \varphi, \psi \rangle \\
&= \langle -A\varphi, \psi \rangle + \lambda_0 \langle \varphi, \psi \rangle \\
&= \langle -(A - \lambda_0)\varphi, \psi \rangle
\end{aligned}$$

is V-elliptic (indeed it is V coercive). We thus find (Theorem 6.1 again or Theorem 6.2) that $A - \lambda_0$, and hence A, is the generator of an analytic semigroup in $H = X = L_2(0, l)$.

6.3   Example 6: The Beam Equation (again)

We return to the beam equation. Recall the system is given by

$$\rho y_{tt} + \gamma y_t + \partial^2 M = f \quad 0 < \xi < l,$$

with

$$y(t, 0) = 0 = \frac{\partial y}{\partial \xi}(t, 0)$$

$$M(t, l) = 0 = \partial M(t, l),$$

where $M(t, \xi) = EI\partial^2 y + c_D I \partial^2 y_t$. We choose as our basic space $H = L_2(0, l)$ with the weighted inner product $\langle \varphi, \psi \rangle_H = \langle \rho\varphi, \psi \rangle_{L_2(0,l)}$. Then the weak form becomes

$$\langle y_{tt} + \frac{\gamma}{\rho} y_t, \varphi \rangle_H + \langle \frac{EI}{\rho} \partial^2 y, \partial^2 \varphi \rangle_H + \langle \frac{c_D I}{\rho} \partial^2 y_t, \partial^2 \varphi \rangle_H = \langle \frac{1}{\rho} f, \varphi \rangle_H$$

for all $\varphi \in V = H_L^2(0,l)$. We choose the weighted inner product for V given by $\langle \varphi, \psi \rangle_V = \int_0^l EI\varphi''\psi''$ (compare the energy inner product of Section 5.2.1).

We define the sesquilinear forms σ_1 and σ_2 on $V \times V \to C$ by

$$\sigma_1(\varphi, \psi) = \langle \frac{EI}{\rho}\varphi'', \psi'' \rangle_H = \int_0^l EI\varphi''\psi''$$

$$\sigma_2(\varphi, \psi) = \langle c_D \frac{I}{\rho}\varphi'', \psi'' \rangle_H + \langle \frac{\gamma}{\rho}\varphi, \psi \rangle_H.$$

The weak form of the equation is then

$$\langle y_{tt}, \varphi \rangle_H + \sigma_1(y, \varphi) + \sigma_2(y_t, \varphi) = \langle \frac{f}{\rho}, \varphi \rangle_H$$

for $\varphi \in V$. To write this in first-order vector form, we use the vector state space $X_E = \mathcal{H} = V \times H$ with the space $\mathcal{V} = V \times V$, noting that $V \hookrightarrow H \hookrightarrow V^*$ and $\mathcal{V} \hookrightarrow \mathcal{H} \hookrightarrow \mathcal{V}^*$ form types of Gelfand triples, where $\mathcal{V}^* = V \times V^*$. Of course, for vector functions $(\varphi_i, \psi_i) \in \mathcal{H}$, i=1,2, we have $\langle (\varphi_1, \psi_1), (\varphi_2, \psi_2) \rangle_{\mathcal{H}} \equiv \langle \varphi_1, \varphi_2 \rangle_V + \langle \psi_1, \psi_2 \rangle_H$.

Homework Exercise

- **Ex. 10** : Explain why we have $\mathcal{V}^* = V \times V^*$ in the Gelfand triple type formulation instead of $\mathcal{V}^* = V^* \times V^*$. (Hint: One cannot make the identification $V \simeq V^*$ when formulating the triple for $\mathcal{V}$ and $\mathcal{H}$ since we have already used $V \subsetneq H \simeq H^* \subsetneq V^*$.)

We define the sesquilinear form $\sigma : \mathcal{V} \times \mathcal{V} \to C$ by (for $\chi = (\varphi, \psi), \zeta = (g, h)$ in $\mathcal{V}$)

$$\sigma(\chi, \zeta) = \sigma((\varphi, \psi), (g, h)) = -\langle \psi, g \rangle_V + \sigma_1(\varphi, h) + \sigma_2(\psi, h).$$

Using the state variable $x(t) = (y(t, \cdot), y_t(t, \cdot))$ in $X_E = \mathcal{H}$, we can rewrite the equation as

$$\langle \dot{x}(t), \chi \rangle_{\mathcal{H}} + \sigma(x(t), \chi) = \langle F(t), \chi \rangle_{\mathcal{H}} \tag{6.6}$$

for $\chi \in \mathcal{V}$, where $F(t) = (0, \frac{1}{\rho}f(t))$.

We readily argue that σ is bounded (continuous) and $\mathcal{V}$-elliptic (actually, $\sigma - \lambda_0| \cdot |_{X_E}^2$ is $\mathcal{V}$-elliptic). Consider first the boundedness argument:

$$|\sigma(\chi, \zeta)| = |\sigma((\varphi, \psi), (g, h))| = |-\langle \psi, g \rangle_V + \sigma_1(\varphi, h) + \sigma_2(\psi, h)|$$

$$\leq |\psi|_V |g|_V + \gamma_1 |\varphi|_V |h|_V + \gamma_2 |\psi|_V |h|_V$$
$$\leq |\chi|_V |\zeta|_V + \gamma_1 |\chi|_V |\zeta|_V + \gamma_2 |\chi|_V |\zeta|_V$$

$$= (1 + \gamma_1 + \gamma_2)|\chi|_V |\zeta|_V$$

for $\chi, \zeta \in \mathcal{V}$. The arguments for $\mathcal{V}$-ellipticity are also simple: for $\chi = (\varphi, \psi) \in \mathcal{V}$ we find

$$\mathrm{Re}\ \sigma(\chi, \chi) = \mathrm{Re}\ \{-\langle \psi, \varphi \rangle_V + \sigma_1(\varphi, \psi) + \sigma_2(\psi, \psi)\}$$

$$= \mathrm{Re}\ \{-\overline{\langle \varphi, \psi \rangle_V} + \langle \varphi, \psi \rangle_V + \sigma_2(\psi, \psi)\}$$

$$= \mathrm{Re}\ \sigma_2(\psi, \psi)$$

$$\geq \delta_2 |\psi|_V^2$$

$$= \delta_2(|\varphi|_V^2 + |\psi|_V^2) - \delta_2 |\varphi|_V^2$$

$$\geq \delta_2(|\varphi|_V^2 + |\psi|_V^2) - \delta_2(|\varphi|_V^2 + |\psi|_H^2)$$

$$= \delta_2 |\chi|_{\mathcal{V}}^2 - \delta_2 |\chi|_{\mathcal{H}}^2.$$

We thus find that $\sigma(\chi, \zeta) = \langle \tilde{\mathcal{A}}\chi, \zeta \rangle_{\mathcal{V}, \mathcal{V}^*}$ gives rise to the infinitesimal generator $\mathcal{A}$ of a C_0 (indeed, analytic–Theorem 6.2) semigroup on $X_E = \mathcal{H}$. It is readily argued that $\sigma(\chi, \zeta) = \langle -\mathcal{A}\chi, \zeta \rangle_{\mathcal{H}}$ for $\chi \in \mathcal{D}(\mathcal{A}) = \{\chi = (\varphi, \psi) \in \mathcal{H} | \psi \in V = H_L^2(0, l), A_1\varphi + A_2\psi \in H, (EI\varphi'' + c_D I\psi'')(l) = 0, (EI\varphi'' + c_D I\psi'')'(l) = 0\}$ where

$$\mathcal{A} = \begin{pmatrix} 0 & I \\ -A_1 & -A_2 \end{pmatrix}$$

with $A_1\varphi = \partial^2(\frac{EI}{\rho}\partial^2\varphi)$ and $A_2\varphi = \partial^2(\frac{c_D I}{\rho}\partial^2\varphi)$.

Thus, the equation (6.6) is the same as

$$\langle \dot{x}(t), \chi \rangle_{\mathcal{H}} + \langle \mathcal{A}x(t), \chi \rangle_{\mathcal{H}} = \langle F(t), \chi \rangle_{\mathcal{H}}$$

which is the appropriate weak sense of

$$\dot{x}(t) = \mathcal{A}x(t) + F(t).$$

This should be compared with the formal formulation in Section 5.2.

Homework Exercises

- Ex. 11 : Some books define $\mathcal{D}(\mathcal{A})$ by

$$\tilde{\mathcal{D}}(\mathcal{A}) = (H^4(0,l) \cap H_L^2(0,l)) \times (H^4(0,l) \cap H_L^2(0,l))$$

  plus boundary conditions. We know $\mathcal{A}|_{\mathcal{D}(\mathcal{A})}$ is an infinitesimal generator of a C_0 semigroup which, in turn, implies $\mathcal{A}|_{\mathcal{D}(\mathcal{A})}$ is a closed operator. You can show $\mathcal{A}|_{\tilde{\mathcal{D}}(\mathcal{A})}$ is not closed. Therefore, we claim that $\mathcal{D}(\mathcal{A}) \neq \tilde{\mathcal{D}}(\mathcal{A})$. Is this true? Look at both the damped and undamped cases.

6.4 Summary of Results on Analytic Semigroup Generation by Sesquilinear Forms

We summarize results available for the special cases of analytic semigroup generation. For further details the reader can consult [BI, T].

Let V and H be complex Hilbert spaces with the Gelfand triple $V \hookrightarrow H \hookrightarrow V^*$. Let $\langle \cdot, \cdot \rangle_{V^*,V}$ be the duality product, and $\sigma : V \times V \to C$ be a sesquilinear form such that σ is

1. V-continuous, i.e., $|\sigma(\varphi, \psi)| \leq \gamma |\varphi|_V |\psi|_V$.

2. V-elliptic, i.e., $\operatorname{Re} \sigma(\varphi, \varphi) \geq \delta |\varphi|_V^2$. (We can, if necessary, replace this by a shift: $\operatorname{Re} \sigma(\varphi, \varphi) + \lambda_0 |\varphi|_H^2 \geq \delta |\varphi|_V^2$.)

As before, let $\hat{A} \in \mathcal{L}(V, V^*)$ (note that this is $-\mathcal{A}$ in our old notation) and $A : D_A \subset H \to H$ be defined such that (see Section 5.6.1)

$$\begin{aligned} \sigma(\varphi, \psi) &= \langle -\hat{A}\varphi, \psi \rangle_{V^*,V} \quad \text{for all } \varphi, \psi \in V \\ &= \langle -A\varphi, \psi \rangle_H \qquad \varphi \in D_A, \psi \in V. \end{aligned}$$

Then from the Lax-Milgram (unbounded form-Theorem 5.2) and the definition of D_A discussed in Section 5.6.1, we have $\mathcal{R}(\hat{A}) = V^*$, $\mathcal{R}(A) = H$, $0 \in \rho(\hat{A})$, and $0 \in \rho(A)$. Indeed we can argue using Theorem 5.2 that $\mathcal{R}(\lambda I - A) = H$ for all $\lambda \geq 0$. We can also note

$$\operatorname{Re} \sigma(\varphi, \varphi) = \operatorname{Re} \langle -\hat{A}\varphi, \varphi \rangle_{V^*,V} \geq \delta |\varphi|_V^2.$$

for all $\varphi \in V$. In other words, $\operatorname{Re}\langle \hat{A}\varphi, \varphi \rangle \leq -\delta |\varphi|_V^2 \leq 0$. Similarly, for $\varphi \in D_A$, $\operatorname{Re} \langle A\varphi, \varphi \rangle_H \leq 0$ which implies A is dissipative. By Lumer-Phillips A is the infinitesimal generator of a C_0 semigroup of contractions $S(t)$ on H. Indeed, by Theorem 6.1 we have that $S(t)$ is an analytic semigroup on H.

We recall the definition (Definition 4.10) of dissipativeness in a Banach space X. An operator $B \in \mathcal{D} \subset X \to X$ is *dissipative* if for each $x \in \mathcal{D}(B)$ there exists $x^* \in F(x) \subset X^*$ such that $\mathrm{Re}\langle x^*, Bx \rangle_{X^*,X} \leq 0$ where $F(x)$ is the duality set. We apply this definition to $X = V^*$, which is a reflexive Banach space in its own right (V is a complex Hilbert space and thus is reflexive and hence so is V^*), with the operator $B = \hat{A}$, $\hat{A} : V \subset V^* \to V^*$. We have $\hat{A}$ being dissipative in the Banach space V^* means for $x \in V$ there exists $x^* \in F(X) \subset X^* = V^{**} = V$ such that $\mathrm{Re}\langle x^*, \hat{A}x \rangle_{V,V^*} \leq 0$ or $\mathrm{Re}\langle \hat{A}x, x^* \rangle_{V^*,V} \leq 0$. However, we have this holding for every $x^* \in V \subset V^*$. (In particular, we can find such a x^* in the duality set.) Therefore, $\hat{A} : V = \mathcal{D}(\hat{A}) \subset V^* \to V^*$ is dissipative. Using Lumer-Phillips again we have $\hat{A}$ is an infinitesimal generator of a C_0 semigroup of contractions $\hat{S}(t)$ on V^* where $\hat{S}(t)|_H = S(t)$.

We thus have that $\hat{A}$ suitably restricted is a generator for semigroups in V^* and H. As we point out in a more detailed remark below, it is of interest in control theory that a suitable restriction of $\hat{A}$ is also a generator of a semigroup in V. Thus it is of practical as well as intellectual interest in having these semigroups (or their suitable restrictions) defined on V, H and V^*. Recall $D_A = \{x \in V | \hat{A}x \in H\}$. We define $\hat{D}_A = \{x \in V | \hat{A}x \in V\}$ and define the operator $\tilde{A} = A|_{\hat{D}_A}$ in V. We can argue that $\mathcal{R}(\tilde{A}) = V$ and moreover the range statement needed for Lumer-Phillips holds for $\tilde{A}$. However, for $\tilde{A}$ to be dissipative in V, we must have for each $x \in \hat{D}_A \subset V$ there exists $x^* \in F(x) \subset V^*$ such that $\mathrm{Re}\langle x^*, \tilde{A}x \rangle_{V^*,V} \leq 0$. We do not directly have that $\tilde{A}$ is dissipative in V. To pursue this further, we need to consider the Tanabe estimates.

6.5　Tanabe Estimates (on "Regular Dissipative Operators")

Suppose a sesquilinear form σ (with associated operator $\hat{A}$) is V-continuous and V-elliptic. Then for $\mathrm{Re}\lambda \geq 0$ and $\lambda \neq 0$, $R_\lambda(\hat{A}) = (\lambda I - \hat{A})^{-1} \in \mathcal{L}(V^*, V)$, and

1. $|R_\lambda(\hat{A})\varphi|_V \leq \frac{1}{\delta}|\varphi|_{V^*}$ for $\varphi \in V^*$. (In other words, $|R_\lambda(\hat{A})|_{\mathcal{L}(V^*,V)} \leq \frac{1}{\delta}$.)

2. $|R_\lambda(\hat{A})\varphi|_H \le \frac{M_0}{|\lambda|}|\varphi|_H$ for $\varphi \in H$ where $M_0 = 1 + \frac{\gamma}{\delta}$. (In other words, $|R_\lambda(\hat{A})|_{\mathcal{L}(H)} \le \frac{M_0}{|\lambda|}$.)

3. $|R_\lambda(\hat{A})\varphi|_{V^*} \le \frac{M_0}{|\lambda|}|\varphi|_{V^*}$ for $\varphi \in V^*$. (In other words, $|R_\lambda(\hat{A})|_{\mathcal{L}(V^*)} \le \frac{M_0}{|\lambda|}$.)

4. $|R_\lambda(\hat{A})\varphi|_V \le \frac{M_0}{|\lambda|}|\varphi|_V$ for $\varphi \in V$. (In other words, $|R_\lambda(\hat{A})|_{\mathcal{L}(V)} \le \frac{M_0}{|\lambda|}$.)

These estimates are discussed more fully in [BI, T]. We only give the arguments for estimate 4, since it is a very strong and useful estimate. Consider the dual or adjoint operator $\hat{A}^* \in \mathcal{L}(V, V^*)$ as defined in Section 4.5. Then we have $\hat{A}^* : V^{**} \simeq V \to V^*$ is given by $\hat{A}^*\psi(\varphi) = \psi(\hat{A}\varphi)$ for $\psi \in V^{**} \simeq V$, $\varphi \in V$. Then

$$
\begin{aligned}
\sigma^*(\psi, \varphi) &= \langle -\hat{A}^*\psi, \varphi \rangle_{V,V^*} = -\hat{A}^*\psi(\varphi) = \psi(-\hat{A}\varphi) \\
&= \langle \psi, -\hat{A}\varphi \rangle_{V,V^*} = \overline{\sigma(\varphi, \psi)}.
\end{aligned}
$$

Thus σ^* is V-continuous and V-elliptic (since σ is) and therefore, $\hat{A}^*$ also satisfies the estimates 1-3 above. Applying 3 to $\hat{A}^*$ we have for Re $\lambda \ge 0$, $\lambda \ne 0$, $\varphi, \psi \in V$

$$
\begin{aligned}
|\langle R_\lambda(\hat{A})\varphi, \psi \rangle_{V,V^*}| &= |\langle \varphi, R_\lambda(\hat{A}^*)\psi \rangle_{V,V^*}| \\
&\le |\varphi|_V |R_\lambda(\hat{A}^*)\psi|_{V^*} \\
&\le |\varphi|_V \frac{M_0}{|\lambda|}|\psi|_{V^*}.
\end{aligned}
$$

Therefore, $|R_\lambda(\hat{A})\varphi|_V \le \frac{M_0}{|\lambda|}|\varphi|_V$.

We can use the Hille-Yosida theorems to show that the C_0 semigroups from above are actually analytic. The theorem below gives a useful condition for analyticity.

Theorem 6.3 *Let $T(t)$ be a C_0 semigroup on a Hilbert space X with infinitesimal generator A, with $0 \in \rho(A)$. Then a semigroup is analytic on X if there exists a constant c such that*

$$
|R_{\mu+i\tau}(A)|_{\mathcal{L}(X)} \le \frac{c}{|\tau|}
$$

for $\mu > 0, \tau \ne 0$ where $\lambda = \mu + i\tau$.

See [P, Theorem II.5.2(b)].

From the Tanabe estimates, we have $|R_\lambda(\hat{A})|_{\mathcal{L}(X)} \le \frac{c}{|\lambda|} = \frac{c}{\sqrt{\mu^2 + \tau^2}} \le \frac{c}{|\tau|}$ for X chosen as $V, H,$ or V^*. Therefore, our estimates suffice to provide analyticity of the associated semigroups. Thus we have the following theorem [BI2].

Theorem 6.4 *Let* $V \hookrightarrow H \hookrightarrow V^*$ *be a Gelfand triple. Assume the sesquilinear form* σ *is* V *continuous and* V-*elliptic. Let* $\hat{A}, A,$ *and* $\hat{\hat{A}}$ *be defined as above. Then*

- $\hat{A}$ *is an infinitesimal generator of an analytic semigroup* $\hat{S}(t)$ *of contractions on* V^*.

- A *is an infinitesimal generator of an analytic semigroup* $S(t)$ *of contractions on* H.

- $\hat{\hat{A}}$ *is an infinitesimal generator of an analytic semigroup* $\hat{\hat{S}}(t)$ *of contractions on* V.

We also have

- $\mathcal{D}_{V^*}(\hat{A}) = V$.

- $\mathcal{D}_H(\hat{A}) = D_A = \{x \in V | \hat{A}x \in H\}$.

- $\mathcal{D}_V(\hat{A}) = \hat{D}_A = \{x \in V | \hat{A}x \in V\}$.

This is usually stated as A or $\hat{A}$ generate an analytic semigroup of contractions on V, H, V^*.

Remark: The results of this section will be of fundamental importance in subsequent discussion on feedback control problems for infinite dimensional systems such as parabolic and strongly damped hyperbolic partial differential equations as well as functional differential equations. These problems can be conveniently and profitably formulated in an abstract setting with the systems defined in terms of Gelfand triples and where the associated algebraic Riccati equations for the feedback gains are formulated in an appropriately defined space V requiring more smoothness than in H. In later sections we will discuss some of the results as developed in [BKcontrol, BI, BI2] and summarized partially in [BSW]. We also remark that in some of these control applications we will have control terms $F(t) = Bu(t)$ in V^* and hence to use the variation of constants representation (1.3) one needs to have the solution semigroup acting in V, H and V^*.

6.6   Infinitesimal Generators in a General Banach Space

Recall that if A is an infinitesimal generator of a C_0 semigroup $T(t)$ in a Hilbert space X, then $S(t) = T^*(t)$ is a C_0 semigroup in X with infinitesimal generator A^*. We have, since A is an infinitesimal generator, $\mathcal{D}(A)$ is dense in X. Thus, if A^* is an infinitesimal generator, $\mathcal{D}(A^*)$ is also dense in X. We can generalize this result in a general Banach space [Pa, Chapter 1].

Theorem 6.5 *If X is a reflexive Banach space and A is an infinitesimal generator of a C_0 semigroup $T(t)$ in X, then A^* is an infinitesimal generator of a C_0 semigroup $S(t)$ in X^* and $S(t) = T^*(t) = (T(t))^*$. In other words, $(e^{A^*t}$ on $X^*)^* = e^{At}$ on X.*

Corollary 6.1 *If $\hat{A}$ is an infinitesimal generator of a C_0 semigroup on V^*, then $\hat{A}^*$ is an infinitesimal generator on $V^{**} = V$ for any reflexive Banach space V.*

In the formulations of the previous sections, we know $\hat{A} \in \mathcal{L}(V, V^*)$ and $\hat{A}^* \in \mathcal{L}(V^{**}, V^*) = \mathcal{L}(V, V^*)$ are infinitesimal generators of C_0 semigroups of contractions on V^*. In other words, $\hat{S}^*(t) = e^{\hat{A}^*t}$ is a C_0 semigroup of contractions on V^*. Applying the previous corollary, we have $(\hat{S}^*(t)$ on $V^*)^* = \hat{S}(t)$ on V. However, $\hat{S}^*(t) \in \mathcal{L}(V^*, V^*)$ implies $(\hat{S}^*(t))^* \in \mathcal{L}(V^{**}, V^{**}) = \mathcal{L}(V, V)$. Since V is a reflexive Hilbert space, $\hat{S}(t)$ defined previously is exactly $\hat{S}(t)|_V$ (here $\hat{A} = A|_{\hat{D}_A}$ is the infinitesimal generator of $\hat{S}(t)$ in V).

The sesquilinear form ideas used in this section to treat first-order (in time) differential equations (recall we converted the second-order beam equation to a first-order vector system before applying the sesquilinear form formulation) can be readily generalized/extended to treat second-order systems directly. We will develop a general second-order theory in subsequent sections. Before doing this, we first discuss the relationship between semigroup solutions and the "weak" solution formulated in terms of sesquilinear formulations for first-order systems.

7 Abstract Cauchy Problems

It is of great practical as well as theoretical interest to know when, and in what sense, solutions of the abstract equations

$$\dot{x}(t) = Ax(t) + f(t)$$
$$x(0) = x_0$$
(7.1)

exist. Moreover, representations of such solutions in terms of a variation of parameters formula and the semigroup generated by A will play a fundamental role in control and estimation formulations. We begin by summarizing results available in the standard literature on linear semigroups and abstract Cauchy problems.

Consider the abstract Cauchy problem (ACP) given by (7.1) where A is the infinitesimal generator of a C_0-semigroup $T(t)$ in a Hilbert space H. We define a *mild solution* x_m of (7.1) as a function in H given by

$$x_m(t) = T(t)x_0 + \int_0^t T(t-s)f(s)ds$$
(7.2)

whenever this entity is well defined (i.e., f is sufficiently smooth).

We say that $x : [0,T] \to H$ is a *strong solution* of (ACP) if $x \in C([0,T],H) \cap C^1((0,T],H)$, $x(t) \in \mathcal{D}(A)$ for $t \in (0,T]$, and x satisfies (7.1) on $[0,T]$.

We have the following series of results from the literature [Li, Pa, W].

Theorem 7.1 *If $f \in L_1((0,T),H)$ and $x_0 \in H$, there is at most one strong solution of (7.1). If a strong solution exists, it is given by (7.2).*

Theorem 7.2 *If $x_0 \in \mathcal{D}(A)$ and $f \in C^1([0,T],H)$, then x_m given by (7.2) provides the unique strong solution of (7.1).*

Theorem 7.3 *If $x_0 \in \mathcal{D}(A), f \in C([0,T],H), f(t) \in \mathcal{D}(A)$ for each $t \in [0,T]$ and $Af \in C([0,T],H)$, then (7.2) provides the unique strong solution of (7.1).*

Theorem 7.4 *Suppose A is the infinitesimal generator of an analytic semigroup $T(t)$ on H. Then if $x_0 \in H$ and f is Hölder continuous (i.e., $|f(t) - f(s)| \le k|t-s|^\gamma$ for some $\gamma \le 1$), then x_m of (7.2) provides the unique strong solution of (7.1).*

Unfortunately, all of these powerful results are too restrictive for use in many applications, including control theory, where typically $f(t) = Bu(t)$ is not continuous, let alone Hölder continuous or C^1. For this reason, a weaker formulation is more appropriate. For this, we follow the presentations of Lions [Li], Wolka [W], and Tanabe [T] which are developed in the context of sesquilinear forms and Gelfand triples, $V \hookrightarrow H \hookrightarrow V^*$, where V, H, V^* are Hilbert spaces.

We define the solution space $\mathcal{W}(0, T)$ by

$$\mathcal{W}(0, T) = \{g \in L_2((0, T), V) : \frac{dg}{dt} \in L_2((0, T), V^*)\}$$

with scalar product

$$\langle g, h \rangle_{\mathcal{W}} = \int_0^T \langle g(t), h(t) \rangle_V dt + \int_0^T \langle \frac{dg}{dt}(t), \frac{dh}{dt}(t) \rangle_{V^*} dt.$$

Then it can be shown (a nice exercise for industrious readers or consult [Li, W]) that $\mathcal{W}(0, T)$ is a Hilbert space which embeds continuously into $C([0, T], H)$.

Assume $\sigma : V \times V \to C$ satisfies for $\varphi, \psi \in V$

$$\text{Re } \sigma(\varphi, \varphi) \geq c_1 |\varphi|_V^2 - \lambda_0 |\varphi|_H^2 \quad c_1 \geq 0, \lambda_0 \text{ real, for all } \varphi \in V,$$

$$|\sigma(\varphi, \psi)| \leq \gamma |\varphi|_V |\psi|_V \quad \text{for all } \varphi, \psi \in V.$$

Then, as already discussed, we have $\mathcal{A} \in \mathcal{L}(V, V^*)$ such that $\sigma(\varphi, \psi) = \langle \mathcal{A}\varphi, \psi \rangle_{V^*,V} = \langle -A\varphi, \psi \rangle_H$ where A is the densely defined restriction of $-\mathcal{A}$ to the set $\mathcal{D}_A = \{\varphi \in V | \mathcal{A}\varphi \in H\}$. We have moreover, that A is the infinitesimal generator of an analytic semigroup $T(t)$ on H. In fact, from Theorem 6.4 we have that $-\mathcal{A}$ is the generator of an analytic semigroup $\mathcal{T}(t)$ in V, H and V^* and $\mathcal{T}(t)$ agrees with $T(t)$ on V and H.

We may consider solutions of (7.1) in the sense of V^*, i.e., in the sense

$$\begin{aligned} \langle \dot{x}(t), \psi \rangle_{V^*,V} + \sigma(x(t), \psi) &= \langle f(t), \psi \rangle_{V^*,V} \quad \text{for } \psi \in V, \\ x(0) &= x_0. \end{aligned} \quad (7.3)$$

By a *strong solution of (7.1) in the V^* sense* (also quite frequently called a *weak* or *variational*—see Chapter 11 for a discussion of the origin of this terminology) or *distributional* solution), we shall mean a function $x = x_{var} \in L_2((0, T), V)$ such that $\dot{x} \in L_2((0, T), V^*)$ and

(7.3) (or equivalently $\dot{x}(t) + \mathcal{A}x(t) = f(t)$) holds almost everywhere on $(0,T)$. Similarly, *mild solutions* $x_m \in V^*$ are given by the (in general, weaker) analogue of (7.2)

$$x_m(t) = \mathcal{T}(t)x_0 + \int_0^t \mathcal{T}(t-s)f(s)ds. \qquad (7.4)$$

We note that this is a generalization of the solutions (7.2) in H to solutions in V^* which of course reduces to (7.2) whenever x_0 and $f(s)$ are in H since $\mathcal{T}(t)$ agrees with $T(t)$ on H. We then have the fundamental existence and uniqueness theorem. (Note that continuous dependence of x_{var} will follow readily from the representation (7.4).)

Theorem 7.5 *Suppose* $x_0 \in H$ *and* $f \in L_2((0,T), V^*)$. *Then (7.3) has a unique strong solution in the* V^* *or variational sense and this is given by the mild solution (7.4).*

Proof We give the arguments which are of interest since they are typical in many finite element method discussions (e.g., see Chapter 12). Let $\{\phi_i\}_1^\infty \subset V$ be a linearly independent total subset (i.e., a basis) of V. We define the "Galerkin" approximations by $x_k(t) = \sum_{i=1}^k w_i(t)\varphi_i$ where the coefficients $\{w_i\}$ are chosen so that

$$\langle \dot{x}_k(t), \varphi_j \rangle_H + \sigma(x_k(t), \varphi_j) = \langle f(t), \varphi_j \rangle_{V^*,V} \qquad (7.5)$$

for $j = 1, \ldots, k$, satisfying the initial condition

$$x_k(0) = x_{k_0}$$

where

$$x_{k_0} = \sum_{i=1}^k w_{i_0}\varphi_i \to x_0$$

in H as $k \to \infty$. Equivalently, (7.5) can be written as

$$\sum_{i=1}^k \dot{w}_i(t)\langle \varphi_i, \varphi_j \rangle + \sum_{i=1}^k w_i(t)\sigma(\varphi_i, \varphi_j) = F_j(t)$$

where $F_j(t) = \langle f(t), \varphi_j \rangle_{V^*,V}$ for $j = 1, \ldots, k$. Therefore, $w_1, \ldots, w_k$ are unique solutions to a vector ordinary differential equation system.

Multiplying (7.5) by w_j and summing over $j = 1, \ldots, k$, we obtain

$$\langle \dot{x}_k(t), x_k(t) \rangle_H + \sigma(x_k(t), x_k(t)) = \langle f(t), x_k(t) \rangle_{V^*, V}$$

with $x_k(0) = x_{k0} \to x_0$ in H. Therefore

$$\frac{1}{2} \frac{d}{dt} |x_k(t)|_H^2 + \sigma(x_k(t), x_k(t)) = \langle f(t), x_k(t) \rangle_{V^*, V}. \tag{7.6}$$

Integrating (7.6), we obtain

$$\frac{1}{2} |x_k(t)|_H^2 - \frac{1}{2} |x_k(0)|_H^2 + \int_0^t \sigma(x_k(s), x_k(s)) ds = \int_0^t \langle f(s), x_k(s) \rangle_{V^*, V} ds.$$

Using the fact that σ is V-elliptic, we have

$$\frac{1}{2} |x_k(t)|_H^2 + c_1 \int_0^t |x_k(s)|_V^2 ds \leq \frac{1}{2} |x_k(0)|_H^2 + \int_0^t |\langle f(s), x_k(s) \rangle_{V^*, V}| ds$$

$$\leq \frac{1}{2} |x_k(0)|_H^2 + \int_0^t (\frac{1}{4\epsilon} |f(s)|_{V^*}^2 + \epsilon |x_k(s)|_V^2) ds$$

Therefore,

$$\frac{1}{2} |x_k(t)|_H^2 + (c_1 - \epsilon) \int_0^t |x_k(s)|_V^2 ds \leq \frac{1}{2} |x_k(0)|_H^2 + \int_0^t \frac{1}{4\epsilon} |f(s)|_{V^*}^2 ds \tag{7.7}$$

or

$$\frac{1}{2} |x_k(t)|_H^2 + (c_1 - \epsilon) \int_0^t |x_k(s)|_V^2 ds \leq \frac{1}{2} |x_k(0)|_H^2 + \frac{1}{4\epsilon} |f|_{L_2((0,t), V^*)}^2.$$

This implies we have $\{x_k\}$ bounded in $C((0, T), H)$ and in $L_2((0, T), V)$. Since $L_2((0, T), V)$ is a Hilbert space, we can choose $\{x_{k_n} | x_{k_n} \rightharpoonup \tilde{x} \in L_2((0, T), V)\}$ to be a convergent subsequence of x_k. Without loss of generality, we reindex and denote x_{k_n} by x_k. Then the limit $\tilde{x}$ is our candidate for a solution where $x_k \rightharpoonup \tilde{x}$ in $L_2((0, T), V)$.

Let $\chi(t) \in C^1(0, T)$ with $\chi(T) = 0$ and $\chi(0) = 0$ and define $\Psi_j(t, \cdot)$ by $\Psi_j(t, \cdot) = \chi(t) \varphi_j$. Multiplying (7.5) by $\chi(t)$ and integrating, we have

$$\int_0^T (\langle \dot{x}_k(t), \varphi_j \rangle_H \chi(t) + \sigma(x_k(t), \varphi_j) \chi(t) - \langle f(t), \varphi_j \rangle_{V^*, V} \chi(t)) \, dt = 0. \tag{7.8}$$

Integrating by parts, we find that (7.8) becomes

$$-\int_0^T \langle x_k(t), \varphi_j \rangle \dot{\chi}(t) dt + \int_0^T \sigma(x_k(t), \varphi_j) \chi(t) - \int_0^T \langle f(t), \varphi_j \rangle_{V^*, V} \chi(t) dt = 0.$$

We can now let $k \to \infty$ and pass the limit through term by term to obtain (this uses the boundedness and continuity properties noted above)

$$\int_0^T -\langle \tilde{x}(t), \varphi_j \rangle \dot{\chi}(t) dt + \int_0^T \sigma(\tilde{x}(t), \varphi_j) \chi(t) dt - \int_0^T \langle f(t), \varphi_j \rangle_{V^*, V} \chi(t) dt = 0,$$

$$(7.9)$$

holding for all $\varphi_j \in V$. Recall that $\{\varphi_j\}$ is a total subset of V and observe that the set of all χ such as chosen above (i.e., $\chi \in C_0^1(0, T)$) are dense in $L_2(0, T)$. Thus we have (7.9) holding for all $\Psi = \varphi \chi$ in $L_2((0, T), V)$. We can rewrite $\sigma(\tilde{x}(t), \varphi) \chi(t)$ as $\mathcal{A}\tilde{x}(t)\Psi$ and $\int_0^T \langle f(t), \varphi \rangle_{V^*, V} \chi(t) dt$ as $f(\Psi)$.

Therefore, (7.9) becomes

$$\langle \frac{d}{dt} \tilde{x}, \Psi \rangle_{V^*, V} + (\mathcal{A}\tilde{x} - f)\Psi = 0$$

where $\Psi \in L_2((0, T), V)$, or $\tilde{x}$ satisfies $\frac{d\tilde{x}}{dt} + \mathcal{A}\tilde{x} - f = 0$ in the $L_2((0, T), V)^*$ sense. However, we have the following needed theorem (details can be found in [E]).

Theorem 7.6 *Let X be a reflexive Banach space. Then*

$$L_p((0, T), X)^* \cong L_q((0, T), X^*)$$

where $\frac{1}{p} + \frac{1}{q} = 1, 1 < p < \infty$.

Returning to our arguments we thus have that the solution to the equation exists in the $L_2((0, T), V)^* = L_2((0, T), V^*)$ sense and is given by $\tilde{x}$. To obtain $\tilde{x}(0) = x_0$, we may use the same arguments with arbitrary $\chi \in C^1(0, T)$, $\chi(T) = 0$, but $\chi(0) \neq 0$.

To prove uniqueness of the solution, it suffices to argue that the solution corresponding to $x_0 = 0$, $f = 0$ is identically zero. With these specific values for f and x_0, (7.3) can be written as

$$\langle \dot{x}(t), \varphi \rangle_{V^*, V} + \sigma(x(t), \varphi) = 0, \qquad (7.10)$$

$$x(0) = 0.$$

Let $\varphi = x(t)$. Then (7.10) becomes

$$\frac{1}{2}\frac{d}{dt}|x(t)|_H^2 + \sigma(x(t), x(t)) = 0.$$

Integrating by parts and using the V-ellipticity of σ, we obtain

$$\frac{1}{2}|x(t)|_H^2 + \int_0^t c_1 |x(s)|_V^2 ds \leq 0.$$

Therefore, $x(t) = 0$ and thus the solution is unique.

To establish continuous dependence of the solution, define

$$x(\cdot; x_0, f) : (x_0, f) \in H \times L_2((0,T), V^*) \rightarrow$$
$$x \in L_2((0,T), V) \bigcap C((0,T), H).$$

Therefore, $x \in L_2((0,T), V^*)$. Taking the limits in (7.7) and using the property that the norms are weakly lower semi-continuous, we obtain the following relation:

$$\frac{1}{2}|x(t)|_H^2 + (c_1 - \epsilon)\int_0^t |x(s)|_V^2 ds \leq \frac{1}{2}|x_0|_H^2 + \frac{1}{4\epsilon}\int_0^t |f(s)|_{V^*}^2 ds,$$

and hence

$$\sup_{t \in [0,T]} \frac{1}{2}|x(t)|_H^2 + (c_1 - \epsilon)\int_0^T |x(s)|_V^2 ds \leq \frac{1}{2}|x_0|_H^2 + \frac{1}{4\epsilon}\int_0^T |f(s)|_{V^*}^2 ds.$$

Since the map $(x_0, f) \rightarrow x$ is a linear map (it is readily seen that solutions of the linear system (7.3) are linear in the initial data and nonhomogeneous forcing function) on $H \times L_2((0,T), V^*)$, we have immediately that x is continuous on $H \times L_2((0,T), V^*)$ into $C((0,T), H)$, $L_2((0,T), V^*)$ and $L_2((0,T), V)$.

Finally, we need to prove the equivalence between this solution and the mild solution given by (7.4). From (7.4) we have that the map $(x_0, f) \rightarrow x(\cdot, x_0, f)$ is continuous from $H \times L_2((0,T), V^*)$ to $L_2((0,T), V^*)$. We have just argued that the variational solution $x_{var}(\cdot, x_0, f)$ is continuous in the same sense. Thus x_{var} and x_m are both continuous in the above sense. Recall that if two continuous functions agree on a dense subset of some set, then the solutions must agree on the entire set. Therefore, if there is a dense subset of $H \times L_2((0,T), V^*)$ on which x_{var} and x_m agree, then they will agree on the entire set.

Choose $x_0 \in D_A$ and $f \in C^1((0,T),H)$. Then Theorem 7.2 guarantees that x_m is the unique solution in the H sense. However, if x_m is a strong solution in the H sense, then it must also be a variational solution (i.e., a strong solution in the V^* sense). However, the mild solution being unique means $x_m(\cdot, x_0, f) = x_{var}(\cdot, x_0, f)$ for $(x_0, f) \in D_A \times C^1((0,T),H)$. But $D_A \times C^1((0,T),H)$ is dense in $H \times L_2((0,T),V^*)$. Hence, we have the equivalence between the solutions corresponding to data (x_0, f) in $H \times L_2((0,T),V^*)$.

We close these discussions on Cauchy problems by remarking that the various types of solutions discussed here and their representations and regularities in V, H or V^* will play an important role in our subsequent discussions of inverse problems as well as in those for the control problems formulated in the latter part of this book.

8 General Second-Order Systems

8.1 Introduction to Second-Order Systems

The ideas in Section 6.3 for Example 6 can be used to treat more general second-order systems. As we shall see in this chapter, the strength of the damping plays a direct role in the regularity properties for the associated semigroup and consequently, the solutions of related equations. Consider the general abstract second-order system

$$\ddot{y}(t) + A_2 \dot{y}(t) + A_1 y(t) = f(t)$$

or, in variational form

$$\langle \ddot{y}(t), \varphi \rangle_{V^*,V} + \sigma_1(y(t), \varphi) + \sigma_2(\dot{y}(t), \varphi) = \langle f(t), \varphi \rangle_{V^*,V} \qquad (8.1)$$

where H is a given complex Hilbert space. As usual, we assume that σ_1 and σ_2 are sesquilinear forms on V where $V \hookrightarrow H \hookrightarrow V^*$ is a Gelfand triple. We also assume that σ_1 is continuous, V-elliptic, and symmetric $(\sigma_1(\varphi, \psi) = \overline{\sigma_1(\psi, \varphi)})$. We assume that σ_2 is continuous and satisfies a weakened ellipticity condition which we formally call H-semiellipticity.

Definition 8.1 A sesquilinear form σ on V is H-*semielliptic* if there is a constant $b \geq 0$ such that

$$\text{Re } \sigma(\varphi, \varphi) \geq b|\varphi|_H^2 \quad \text{for all } v \in V.$$

Note that $b = 0$ is allowed in this definition.

Since σ_1 and σ_2 are continuous, we have that there exists $A_i \in \mathcal{L}(V, V^*)$, $i = 1, 2$, such that

$$\sigma_i(\varphi, \psi) = \langle A_i \varphi, \psi \rangle_{V^*,V} \quad \text{for all } \varphi, \psi \in V, \quad i = 1, 2.$$

Following the ideas in Section 6.3 for Example 6, we define spaces $\mathcal{V} = V \times V$ and $\mathcal{H} = V \times H$ and rewrite our second-order system as a first-order vector system. Defining, for $\chi = (\varphi, \psi), \zeta = (g, h) \in \mathcal{V}$, the sesquilinear form

$$\sigma(\chi, \zeta) = \sigma((\varphi, \psi), (g, h)) = -\langle \psi, g \rangle_V + \sigma_1(\varphi, h) + \sigma_2(\psi, h),$$

we can write our system for $x(t) = (y(t), \dot{y}(t))$ as

$$\langle \dot{x}(t), \chi \rangle_{\mathcal{H}} + \sigma(x(t), \chi) = \langle F(t), \chi \rangle_{\mathcal{H}} \quad \chi \in V$$

where $F(t) = (0, f(t))$. This is formally equivalent to the system

$$\dot{x}(t) = \mathcal{A}x(t) + F(t)$$

where $\mathcal{A}$ is given by

$$\mathcal{D}(\mathcal{A}) = \{x = (\varphi, \psi) \in \mathcal{H} | \psi \in V \text{ and } A_1\varphi + A_2\psi \in H\} \qquad (8.2)$$

and

$$\mathcal{A} = \begin{pmatrix} 0 & I \\ -A_1 & -A_2 \end{pmatrix}. \qquad (8.3)$$

We first note that σ is V continuous. To see this, we observe that σ_1 and σ_2 being V continuous implies

$$\sigma_1(\varphi, h) \le \gamma_1 |\varphi|_V |h|_V$$

and

$$\sigma_2(\varphi, h) \le \gamma_2 |\psi|_V |h|_V.$$

We also have $|\chi|_V^2 = |\varphi|_V^2 + |\psi|_V^2$ and $|\zeta|_V^2 = |g|_V^2 + |h|_V^2$. Putting all of this together, we have from standard arguments given in the previous chapter

$$\begin{aligned} |\sigma((\varphi, \psi), (g, h))| &\le |\psi|_V |g|_V + \gamma_1 |\varphi|_V |h|_V + \gamma_2 |\psi|_V |h|_V \\ &\le |\chi|_V |\zeta|_V + \gamma_1 |\chi|_V |\zeta|_V + \gamma_2 |\chi|_V |\zeta|_V \\ &= (1 + \gamma_1 + \gamma_2)|\chi|_V |\zeta|_V. \end{aligned}$$

This indeed implies that σ is V continuous.

As σ is V-continuous, $\mathcal{A}$ is the negative of the restriction to $\mathcal{D}(\mathcal{A})$ of the operator $\tilde{\mathcal{A}} \in \mathcal{L}(V, V^*)$ defined by $\sigma(\chi, \zeta) = \langle \tilde{\mathcal{A}}\chi, \zeta \rangle_{V^*, V}$ so that $\sigma(\chi, \zeta) = \langle -\mathcal{A}\chi, \zeta \rangle_{\mathcal{H}}$ for $\chi \in \mathcal{D}(\mathcal{A}), \zeta \in V$ (recall the formulations of Section 5.2).

8.2   Results for σ_2 V-elliptic

If both σ_1 and σ_2 are V-elliptic and σ_1 is the same as the V inner product, then we have exactly the case of Kelvin-Voigt damping in Example 6. We proved with these assumptions, σ is $\mathcal{V}$-elliptic. (Actually, we proved $\sigma(\cdot, \cdot) + \lambda_0 \langle \cdot, \cdot \rangle_{\mathcal{H}}$ is $\mathcal{V}$-elliptic.) Therefore, we have $\mathcal{A}$ is the infinitesimal generator of an analytic semigroup (not of contractions since $\lambda_0 > 0$) on $\mathcal{H}$.

Even if the V inner product and σ_1 are not the same, this result is true. Since σ_1 is continuous, we have $|\sigma_1(\varphi, \varphi)| \leq \gamma_1 |\varphi|_V^2$ while σ_1 is symmetric (i.e., $\sigma_1(\varphi, \psi) = \overline{\sigma_1(\psi, \varphi)}$) implies Re $\sigma_1(\varphi, \varphi) = \sigma_1(\varphi, \varphi)$. Thus, σ_1 being V-elliptic is equivalent to σ_1 being V-coercive: $\sigma_1(\varphi, \varphi) \geq \delta |\varphi|_V^2$. Hence, σ_1 and the inner product are topologically equivalent. We may thus define V_1 as the space V with σ_1 as inner product, obtaining a space that is setwise equal and topologically equivalent to V. In the space $\mathcal{H}_1 = V_1 \times H$ the operator $\mathcal{A}$ is now associated with the $\mathcal{V}_1$-elliptic (where $\mathcal{V}_1 = V_1 \times V_1$) form $\sigma^{(1)}(\chi, \zeta) = \langle -\mathcal{A}\chi, \zeta \rangle_{\mathcal{H}_1}$ that (as we argued in Section 6.3) satisfies the conditions of our Theorem 6.2. Hence, $\mathcal{A}$ generates an analytic semigroup in $\mathcal{H}_1$ and hence an analytic semigroup in the equivalent space $\mathcal{H}$. Thus we have

Theorem 8.1 *Let $V \hookrightarrow H \hookrightarrow V^*$ and suppose that σ_1 and σ_2 of (8.1) are V-continuous and V-elliptic sesquilinear forms on V and that σ_1 is symmetric. Then the operator $\mathcal{A}$ defined in (8.2) and (8.3) is the infinitesimal generator of an analytic semigroup in $\mathcal{H} = V \times H$.*

Remark 2 *By the results of Section 6.4, we actually have under the assumptions above that $\mathcal{A}$ generates an analytic semigroup on $\mathcal{V}$, $\mathcal{H}$ and $\mathcal{V}^*$.*

8.3 Results for σ_2 H-semielliptic

If σ_2 is not V-elliptic, then we will not, in general, obtain an analytic solution semigroup for our system. We will obtain a C_0 semigroup, but must work a little more to obtain such. So assume that σ_2 is only H-semielliptic. Then we have $\mathcal{A}$ defined in (8.2) and (8.3) is dissipative in $\mathcal{H}_1$ since

$$
\begin{aligned}
\text{Re} \, \langle \mathcal{A}x, x \rangle_{\mathcal{H}_1} &= \text{Re} \, \{\sigma_1(\psi, \varphi) - \sigma_1(\varphi, \psi) - \sigma_2(\psi, \psi)\} \\
&= \text{Re} \, \{\overline{\sigma_1(\varphi, \psi)} - \sigma_1(\varphi, \psi) - \sigma_2(\psi, \psi)\} \\
&= -\text{Re} \, \sigma_2(\psi, \psi) \\
&\leq -b|\psi|_H^2 \leq 0.
\end{aligned}
$$

To argue that $\mathcal{A}$ is a generator, we use the Lumer-Phillips theorem; thus we need to argue that for some $\lambda > 0$, the range of $\lambda I - \mathcal{A}$ is $\mathcal{H}_1$. Given $\zeta = (g, h) \in \mathcal{H}_1$, we wish to solve $(\lambda - \mathcal{A})\chi = \zeta$ for $\chi = (\varphi, \psi) \in \mathcal{D}(\mathcal{A})$.

So we consider the equation

$$(\lambda - \mathcal{A})(\varphi, \psi) = (g, h) \quad \text{for } (g, h) \in V_1 \times H.$$

This is equivalent to

$$\begin{cases} \lambda\varphi - \psi & = g \\ \lambda\psi + A_1\varphi + A_2\psi & = h. \end{cases} \tag{8.4}$$

If we formally solve the first equation for $\psi = \lambda\varphi - g$ and substitute this into the second equation, we obtain

$$\lambda^2\varphi - \lambda g + A_1\varphi + A_2(\lambda\varphi - g) = h$$

or

$$\lambda^2\varphi + A_1\varphi + \lambda A_2\varphi = h + \lambda g + A_2 g. \tag{8.5}$$

This equation must be solved for $\varphi \in V_1$ (and then ψ defined by $\psi = \lambda\varphi - g$ will also be in V_1).

These formal calculations suggest that we define for $\lambda > 0$ the associated sesquilinear form on $V \times V \to C$

$$\sigma_\lambda(\varphi, \psi) = \lambda^2 \langle \varphi, \psi \rangle_H + \sigma_1(\varphi, \psi) + \lambda\sigma_2(\varphi, \psi).$$

Since σ_1 is V-elliptic and σ_2 is H-semielliptic we have

$$\begin{aligned} \text{Re } \sigma_\lambda(\varphi, \varphi) &= \lambda^2 |\varphi|_H^2 + \text{Re } \sigma_1(\varphi, \varphi) + \lambda \text{ Re } \sigma_2(\varphi, \varphi) \\ &\geq \lambda^2 |\varphi|_H^2 + c_1 |\varphi|_V^2 + \lambda b |\varphi|_H^2 \\ &= \lambda(\lambda + b)|\varphi|_H^2 + c_1 |\varphi|_V^2 \\ &> c_1 |\varphi|_V^2 \end{aligned}$$

for $\tilde{\lambda} = \lambda(\lambda + b) > 0$. Hence σ_λ is V-elliptic and (8.5) is solvable for $\varphi \in V$ by Lax-Milgram. It follows that (8.4) is solvable for $(\varphi, \psi) \in \mathcal{D}(\mathcal{A})$, i.e., $\mathcal{R}(\lambda - \mathcal{A}) = \mathcal{H}_1$. Thus we have that $\mathcal{A}$ generates a contraction semigroup in $\mathcal{H}_1$ and a C_0 semigroup in $\mathcal{H}$.

Theorem 8.2 *Let $V \hookrightarrow H \hookrightarrow V^*$ and suppose that σ_1 and σ_2 of (8.1) satisfy: σ_1 is V-elliptic, V continuous and symmetric, σ_2 is V continuous and H-semielliptic. Then $\mathcal{A}$ defined by (8.2) and (8.3) generates a C_0 semigroup in $\mathcal{H}$.*

8.4 Stronger Assumptions for σ_2

If we strengthen the assumption on the damping form, we can obtain a stronger result.

Theorem 8.3 *Suppose σ_1 is V-elliptic, V continuous, and symmetric and σ_2 is H-elliptic, V continuous, and symmetric. Then $\mathcal{A}$ is the infinitesimal generator of a C_0 semigroup $T(t)$ in $\mathcal{H} = V \times H$ that is exponentially stable, i.e., $|T(t)\chi|_\mathcal{H} \leq Me^{-wt}|\chi|_\mathcal{H}$ for some $w > 0$.*

To motivate the arguments used to establish this result, we consider for $w > 0$ the change of dependent variable $y(t) = e^{-wt}r(t)$ in the equation

$$\ddot{y} + A_2\dot{y}(t) + A_1 y(t) = 0. \tag{8.6}$$

Upon substitution, we obtain

$$\ddot{r}(t) + \hat{A}_2\dot{r}(t) + \hat{A}_1 r(t) = 0 \tag{8.7}$$

where

$$\hat{A}_1 = A_1 - wA_2 + w^2 I$$

$$\hat{A}_2 = A_2 - 2wI.$$

This suggests that we define the sesquilinear forms

$$\hat{\sigma}_1(\varphi, \psi) = \sigma_1(\varphi, \psi) - w\sigma_2(\varphi, \psi) + w^2\langle\varphi, \psi\rangle_H$$

$$\hat{\sigma}_2(\varphi, \psi) = \sigma_2(\varphi, \psi) - 2w\langle\varphi, \psi\rangle_H$$

so that $\hat{\sigma}_i(\varphi, \psi) = \langle\hat{A}_i\varphi, \psi\rangle_{V^*,V}, i = 1, 2$, and the transformed variational form of (8.6) is

$$\langle\ddot{r}(t), \varphi\rangle_{V^*,V} + \hat{\sigma}_1(r(t), \varphi) + \hat{\sigma}_2(\dot{r}(t), \varphi) = 0$$

for $\varphi \in V$.

We observe that $\hat{\sigma}_1, \hat{\sigma}_2$ are continuous and $\hat{\sigma}_1$ is symmetric since both σ_1 and σ_2 are. Since σ_2 is symmetric (hence $\sigma_2(\varphi, \varphi)$ is real) and continuous with $\sigma_2(\varphi, \varphi) \leq k_2|\varphi|_V^2$, we have for $\varphi \in V$

$$
\begin{aligned}
\text{Re } \hat{\sigma}_1(\varphi, \varphi) &= \hat{\sigma}_1(\varphi, \varphi) \\
&= \sigma_1(\varphi, \varphi) - w\sigma_2(\varphi, \varphi) + w^2|\varphi|_H^2 \\
&\geq c_1|\varphi|_V^2 - w\gamma_2|\varphi|_V^2 + w^2|\varphi|_H^2 \\
&\geq (c_1 - w\gamma_2)|\varphi|_V^2.
\end{aligned}
$$

Hence $\hat{\sigma}_1$ is V-elliptic if $w > 0$ is chosen so that $w < \frac{c_1}{\gamma_2}$.

Moreover, we find that $\hat{\sigma}_2$ is H-semielliptic if w is chosen properly since

$$\text{Re } \hat{\sigma}_2(\varphi, \varphi) = \text{Re } \sigma_2(\varphi, \varphi) - 2w|\varphi|_H^2 \geq (b - 2w)|\varphi|_H^2.$$

Therefore, $\hat{\sigma}_2$ is H-semielliptic if $w < \frac{b}{2}$.

Thus, if we choose $w > 0$ as $w = \frac{1}{2} \min \{\frac{b}{2}, \frac{c_1}{\gamma_2}\}$, we find that $\hat{\sigma}_1$ and $\hat{\sigma}_2$ satisfy the assumptions of Theorem 8.2. By the arguments preceding that theorem, we see that

$$\hat{A} = \begin{pmatrix} 0 & I \\ -\hat{A}_1 & -\hat{A}_2 \end{pmatrix}$$

(see (8.2) and (8.3)) generates a contraction semigroup $\hat{T}(t)$ on $\hat{\mathcal{H}}_1 = \hat{V}_1 \times H$ where $\hat{V}_1$ is V taken with $\hat{\sigma}_1$ as the inner product ($\hat{V}_1$ is topologically equivalent to V).

Now let $T(t)$ be the C_0-semigroup generated by $\mathcal{A}$ (see (8.2), (8.3) and Theorem 8.2). If $x(t) = \begin{pmatrix} y(t) \\ \dot{y}(t) \end{pmatrix}$ and $w(t) = \begin{pmatrix} r(t) \\ \dot{r}(t) \end{pmatrix}$ are solutions of (8.6) and (8.7) respectively, we have $x(t) = T(t)x_0$ where $x_0 = \begin{pmatrix} y_0 \\ w_0 \end{pmatrix}$ and $w(t) = \hat{T}(t)w_0$. Since $y(t) = e^{-wt}r(t)$ and $\dot{y}(t) = -we^{wt}r(t) + e^{wt}\dot{r}(t)$, we see that $x(t) = e^{wt}\Gamma w(t)$ where

$$\Gamma = \begin{pmatrix} 1 & 0 \\ -w & 1 \end{pmatrix}$$

and $w_0 = \Gamma^{-1}x_0$. It follows since $|\hat{T}(t)|_{\hat{\mathcal{H}}_1} \leq 1$ that

$$|T(t)x_0|_{\hat{\mathcal{H}}_1} \leq e^{-wt}|\Gamma\hat{T}(t)\Gamma^{-1}x_0|_{\hat{\mathcal{H}}_1}$$

$$\leq Me^{-wt}|x_0|_{\hat{\mathcal{H}}_1}.$$

Since $\hat{\mathcal{H}}_1$ and $\mathcal{H} = V \times H$ are norm equivalent, we thus find that the semigroup $T(t)$ is exponentially stable in $\mathcal{H}$.

We note again that results such as these (exponential stability of the underlying semigroup) will play a fundamental role in our control theory discussions of Chapters 16 and 17.

9 Weak Formulations for Second-Order Systems

Our focus on dynamical systems in this book to this point has involved semigroup related solutions and representations. We also introduced sesquilinear forms in the context of semigroup generation and certain weak forms of differential equations. In numerous applications (some of which are discussed below) the coefficients in the dynamical systems are dependent on time. The semigroup approach of earlier chapters is not directly useful in such problems. Even though there are generalizations of semigroups involving the theory of evolution operators [Amann, DaPrato, Goldstein, Kreĭn, Martin] to which one could appeal, we will not discuss that in this book. However, generalizations of the sesquilinear form approach discussed earlier can be made to include time dependent systems. We present such an approach in this chapter.

9.1 Model Formulation

We return to the general second-order systems (8.1) of Chapter 8 to illustrate well-posedness ideas in the context of the abstract hyperbolic model with time dependent stiffness and damping given by

$$\langle \ddot{y}(t), \psi \rangle_{V^*,V} + d(t; \dot{y}(t), \psi) + a(t; y(t), \psi) = \langle f(t), \psi \rangle_{V^*,V}$$

where $V \subset V_D \subset H \subset V_D^* \subset V^*$ are Hilbert spaces with continuous and dense injections, where H is identified with its dual and $\langle \cdot, \cdot \rangle$ denotes the associated duality product. We first show under reasonable assumptions on the time-dependent sesquilinear forms $a(t; \cdot, \cdot) : V \times V \rightarrow C$ and $d(t; \cdot, \cdot) : V_D \times V_D \rightarrow C$ that this model possesses a unique solution and that the solution depends continuously on the data of the problem. We also consider well-posedness as well as finite element type approximations in associated inverse problems. We treat a weak formulation in abstract differential operator form that includes plate, beam, and shell equations with several important kinds of damping [BSW]. A quick remark on notation in this chapter: we will use the notation involving $d(t; \cdot, \cdot)$ on $V_D \times V_D$ in place of $\sigma_2(\cdot, \cdot)$ on $V \times V$. We do this to allow a separate space V_D for damping (which permits weaker damping) to emphasize the similar, but separate roles that V and V_D play, especially in the time dependent case for damping $d(t; \cdot, \cdot)$ which may in some cases actually be defined on $H \times H$.

Applications for such systems are abundant and range from systems with periodic or structured pattern time dependences that occur for example in thermally dependent systems orbiting in space (periodic exposure to sunlight) to earth bound structures (bridges, buildings) and aircraft/space structure components with extreme temperature exposures (winter vs. summer). On a longer time scale, such systems are important in time dependent health of elastic structures where long term (slowly varying) time dependence of parameters may be used in detecting aging/fatiguing as represented by changes in stiffness and/or damping. Moreover, applications may be found in modern "smart material structures" [BSW, RCS] where one "controls" damping and/or elasticity via piezoceramic patches, electrically active polymers, etc.

As before let V and H be complex Hilbert spaces forming a Gelfand triple $V \subset H \cong H^* \subset V^*$ with duality product $\langle \cdot, \cdot \rangle_{V^*,V}$. The injections are assumed to be dense and continuous and the spaces are assumed to be separable. Moreover, we assume that there exists a Hilbert space V_D (the damping space), such that $V \subset V_D \subset H = H^* \subset V_D^* \subset V^*$, allowing for a wide class of damping models. Thus the duality products $\langle \cdot, \cdot \rangle_{V^*,V}$ and $\langle \cdot, \cdot \rangle_{V_D^*,V_D}$ are the natural extensions by continuity of the inner product $\langle \cdot, \cdot \rangle$ in H to $V^* \times V$ and $V_D^* \times V_D$, respectively. The H-norm is denoted by $|\cdot|$ or $|\cdot|_H$ for clarity if needed and the V and V_D-norms are denoted $|\cdot|_V$ and $|\cdot|_{V_D}$, respectively. We denote $\partial_t y$ by $\dot{y}$, where $y = y(t)$ is considered as a function of time taking values in one of the occurring Hilbert spaces, that is, $y(t,\xi) = y(t)(\xi)$ where $y(t) \in H$, for example.

Our goal is to investigate the system

$$\ddot{y}(t) + D(t)\dot{y}(t) + A(t)y(t) = f(t), \quad \text{in } V^*, \quad t \in (0,T), \qquad (9.1)$$

with $y(0) = y^0 \in V$ and $\dot{y}(0) = y^1 \in H$ which is equivalent to

$$\langle \ddot{y}(t), \psi \rangle_{V^*,V} + \langle D(t)\dot{y}(t), \psi \rangle_{V_D^*,V_D} + \langle A(t)y(t), \psi \rangle_{V^*,V} = \langle f(t), \psi \rangle_{V^*,V},$$
$$(9.2)$$

for all $\psi \in V$ and $t \in (0,T)$, with $y(0) = y^0 \in V$ and $\dot{y}(0) = y^1 \in H$.

Following standard terminology we will call a function $y \in L_2(0,T;V)$, with $\dot{y} \in L_2(0,T;H)$ and $\ddot{y} \in L_2(0,T;V^*)$ a *weak solution* of the initial-value problem (9.1) if it solves the equation (9.2) in the strong V^* sense of Chapter 7, or, equivalently, solves the equation (9.11) below, with $y(0) = y^0 \in V$ and $\dot{y}(0) = y^1 \in H$ given.

We will assume that the operators A and D arise (as defined precisely in (9.8)-(9.9) below) from time-dependent sesquilinear forms a

and d satisfying the following natural ellipticity, coercivity, and differentiability conditions.

First, we assume hermitian symmetry, that is

(H1) $a(t, \phi, \psi) = \overline{a(t, \psi, \phi)}$ for all $\phi, \psi \in V$.

(H2) $|a(t; \phi, \psi)| \leq c_1 |\phi|_V |\psi|_V$, $\phi, \psi \in V$ where c_1 is independent of t.

We assume, further, that

(H3) $a(t; \phi, \psi)$ for $\phi, \psi \in V$ fixed is *continuously differentiable* with respect to t for $t \in [0, T]$ (T finite) and

$$|\dot{a}(t; \phi, \psi)| \leq c_2 |\phi|_V |\psi|_V, \quad \text{for all } t \in [0, T], \qquad (9.3)$$

with c_2 once again independent of t.

We also assume that the sesquilinear form $a(t; \phi, \psi)$ is *V-elliptic*, so that

(H4) There exists a constant $\alpha > 0$ such that

$$|a(t; \phi, \phi)| \geq \alpha |\phi|_V^2 \text{ for all } t \in [0, T] \text{ and for all } \phi \in V. \qquad (9.4)$$

For the sesquilinear form d we assume similarly

(H5) $|d(t; \phi, \psi)| \leq c_3 |\phi|_{V_D} |\psi|_{V_D}$, $\phi, \psi \in V_D$ where c_3 is independent of t.

We assume, further, that

(H6) $d(t; \phi, \psi)$ for $\phi, \psi \in V_D$ fixed is *continuously differentiable* with respect to t for $t \in [0, T]$ (T finite) and

$$|\dot{d}(t; \phi, \psi)| \leq c_4 |\phi|_{V_D} |\psi|_{V_D}, \quad \text{for all } t \in [0, T], \qquad (9.5)$$

c_4 once again independent of t.

Then $t \to d(t; \phi, \psi)$ and $t \to a(t; \phi, \psi)$ are $C^1[0, T]$ for all $\phi, \psi \in V_D$, and $\phi, \psi \in V$, respectively, which implies that $d(t; \phi, \psi)$ and $a(t; \phi, \psi)$ are sufficiently well-behaved in order to have existence for (9.1) or (9.2). We also assume that the sesquilinear form $d(t; \phi, \psi)$ is V_D-coercive. That is,

(H7) There exist constants λ_d and $\alpha_d > 0$, such that

$$\operatorname{Re} d(t; \phi, \phi) + \lambda_d |\phi|^2 \geq \alpha_d |\phi|_{V_D}^2 \text{ for all } t \in [0, T] \text{ and for all } \phi \in V_D. \tag{9.6}$$

We know then from our earlier considerations (Chapter 8) that there exist representation operators $A(t)$ and $D(t)$

$$A(t) : V \to V^*, \quad D(t) : V_D \to V_D^*, \tag{9.7}$$

which for each fixed t are continuous and linear, with

$$a(t; \phi, \psi) = \langle A(t)\phi, \psi \rangle_{V^*, V}, \quad \text{for all} \quad \phi, \psi \in V, \tag{9.8}$$

and

$$d(t; \phi, \psi) = \langle D(t)\phi, \psi \rangle_{V_D^*, V_D}, \quad \text{for all} \quad \phi, \psi \in V_D. \tag{9.9}$$

We will now consider the following problem: Given finite T and $f \in L_2(0, T; V_D^*)$ along with initial conditions

$$y^0 \in V, \quad y^1 \in H,$$

we wish to find a function $y \in L_2(0, T; V), \dot{y} \in L_2(0, T; V_D)$ such that in V^* we have

$$\begin{cases} \ddot{y}(t) + D(t)\dot{y}(t) + A(t)y(t) = f(t), & t \in (0, T), \\ y(0) = y^0, \ \dot{y}(0) = y^1. \end{cases} \tag{9.10}$$

That is, for $f \in L_2(0, T; V_D^*)$

$$\langle \ddot{y}(t), \psi \rangle_{V^*, V} + d(t; \dot{y}(t), \psi) + a(t; y(t), \psi) = \langle f(t), \psi \rangle_{V^*, V} \quad \text{for all } \psi \in V. \tag{9.11}$$

We remark that (9.11) is meaningful since $f(t) \in V_D^* \subset V^*$.

This formulation covers linear beam (i.e., Example 6 discussed earlier), plate and shell models with numerous damping models (Kelvin-Voigt, viscous, square-root, structural, and spatial hysteresis) frequently studied in the literature. The formulation above is non-standard in the sense that the damping sesquilinear form is incorporated in the variational model *and* is time-dependent. The problem without damping $(d = 0)$ and $f \in L_2(0, T; H)$ was treated by Lions in [Li], and subsequently in [W]. The less general case without damping and $V = H_0^1(\Omega)$ is treated for example in [Ev]. The model above with $d : V_D \to C$ independent of time is treated in [BSW]. The following theorem is a time-dependent extension of the previous results.

Theorem 9.1 *Assume* $(f, y^0, y^1) \in L_2(0, T; V_D^*) \times V \times H$ *and that conditions* **(H1)-(H7)** *hold. Then there exists a unique solution* y *to (9.11) with* $(y, \dot{y}) \in L_2(0, T; V) \times L_2(0, T; V_D)$, *and the mapping*

$$(f, y^0, y^1) \to (y, \dot{y}), \tag{9.12}$$

is continuous and linear on

$$L_2(0, T; V_D^*) \times V \times H \to L_2(0, T; V) \times L_2(0, T; V_D). \tag{9.13}$$

As we will see, Theorem 9.1 can then be extended to stronger smoothness results on solutions.

Theorem 9.2 *Assume that* $(f, y^0, y^1) \in L_2(0, T; V_D^*) \times V \times H$ *and that conditions* **(H1)-(H7)** *hold. Then there exists (perhaps after modifications on a set of measure zero) a unique solution* y *to (9.11) with* $(y, \dot{y}) \in C(0, T; V) \times (C(0, T; H) \cap L_2(0, T; V_D))$, *and the mapping*

$$(f, y^0, y^1) \to (y, \dot{y}), \tag{9.14}$$

is continuous and linear on

$$L_2(0, T; V_D^*) \times V \times H \to C(0, T; V) \times (C(0, T; H) \cap L_2(0, T; V_D)). \tag{9.15}$$

We shall present arguments for these results below, after some remarks and discussion of possible underlying models.

Remark: If we only have that the inequality (9.6) for the damping sesquilinear form d is satisfied with $\alpha_d = 0$, the results are still true with modifications. Then it will be necessary that $f \in L_2(0, T; H)$ and one obtains only that $\dot{y} \in L_2(0, T; H)$; that is, we have the same results as if there was no damping. (See, e.g., [Li].) As an added comment, we note that J. L. Lions was one of the early and most prolific contributors to functional analytic formulations of partial differential equations and control theory. His many contributions include the seminal texts [Li, LiMag].

9.2 Discussion of the Model

It is well known from the literature that the strong form of the operator formulation (9.1) of the problem in general causes computational problems due to irregularities stemming from non-smooth terms typically in the force/moment terms in, for example, elasticity problems.

The weak formulation has proven advantageous both for theoretical and practical purposes, specifically in the effort to estimate parameters or for control purposes [BSW]. To give a particular example illustrating and motivating our discussions here, we consider a version of the beam of Example 6 but with both ends fixed. For an Euler-Bernoulli beam of length l, width b, thickness h, and linear density ρ, where the parameters b, h, ρ may be functions that depend on time and/or spatial position ξ along the beam, the equation for transverse displacements $y = y(t, \xi)$ (in strong form [BSW]) is given by

$$\rho\frac{\partial^2 y}{\partial t^2} + \frac{\partial^2}{\partial \xi^2}\left\{\widetilde{C_D I}\frac{\partial^3 y}{\partial \xi^2 \partial t} + \widetilde{EI}\frac{\partial^2 y}{\partial \xi^2}\right\} = f \quad 0 < \xi < l, \qquad (9.16)$$

with fixed end boundary conditions

$$y(t, 0) = \frac{\partial y}{\partial \xi}(t, 0) = y(t, l) = \frac{\partial y}{\partial \xi}(t, l) = 0. \qquad (9.17)$$

Here we assume Kelvin-Voigt structural damping with damping coefficient $\widetilde{C_D I} = \widetilde{C_D I}(t, \xi)$ and the possibly time and spatially dependent stiffness coefficient given by $\widetilde{EI} = \widetilde{EI}(t, \xi)$. For simplicity we assume ρ is constant and scale the system by taking $\rho = 1$. One can readily compute that $\widetilde{EI} = Eh^3 b/12, \widetilde{C_D I} = C_D h^3 b/12$, where the Young's modulus E, the damping coefficient C_D, and the geometric parameters h, b may in general all be time and/or spatially dependent. As we have seen, in weak form this can be written

$$\langle \ddot{y}(t), \psi \rangle_{V^*, V} + \langle \widetilde{C_D I}\frac{\partial^2 \dot{y}(t)}{\partial \xi^2} + \widetilde{EI}\frac{\partial^2 y(t)}{\partial \xi^2}, \frac{\partial^2 \psi}{\partial \xi^2}\rangle_H = \langle f(t), \psi \rangle_{V^*, V},$$
$$\qquad (9.18)$$

for all $\psi \in V$, where $H = L_2(0, l)$ and $V = V_D = H_0^2(0, l)$ with

$$H_0^2(0, l) \equiv \{\psi \in H^2(0, l)|\psi(0) = \psi'(0) = \psi(l) = \psi'(l) = 0\}. \qquad (9.19)$$

Here we have adopted the usual notation $\psi' = \frac{\partial \psi}{\partial \xi}$.

This has the form (9.2) or (9.11) with

$$a(t; \phi, \psi) = \langle A(t)\phi, \psi \rangle_{V^*, V} = \int_0^l \widetilde{EI}(t, \xi)\phi''(\xi)\psi''(\xi)d\xi \qquad (9.20)$$

$$d(t; \phi, \psi) = \langle D(t)\phi, \psi \rangle_{V^*, V} = \int_0^l \widetilde{C_D I}(t, \xi)\phi''(\xi)\psi''(\xi)d\xi. \qquad (9.21)$$

Models such as this can be generalized to higher dimensions (in $\mathbb{R}^n$ for $n = 2, 3$) to treat more general beams, plates, shells, and solid bodies [BSW, RCS]. Of great interest are a number of useful damping models that can be readily used with these equations and treated using the abstract formulation developed here. We discuss briefly some of these damping models without going into much detail, as the purpose of this section is to establish well-posedness and approximation properties of the abstract model. We consider briefly several damping models of interest in practice [BFW88, BFW89, BFWIC, BaIn, BaPed, BSW, BWIC].

Time-dependent Kelvin-Voigt damping: Let $\omega \subset [0, l]$, with 1_ω denoting the characteristic function of ω. Let $\gamma, \delta > 0$ denote material parameters and let $t \to k(t)$ denote a sufficiently smooth function. Then a *time-dependent* damping sesquilinear form is given by

$$d(t; \phi, \psi) = \int_0^l (\gamma + \delta k(t) 1_\omega(\xi)) \phi''(\xi) \psi''(\xi) d\xi, \qquad (9.22)$$

for $\phi, \psi \in V_D = V = H_0^2(0, l)$. This gives a model for a mechanical structure damped by a time-varying actuator, localized somewhere inside the structure. This could be piezoceramic actuators or other "smart" devices, with the possibility of them varying in time.

Time-dependent viscous damping: This is a velocity-proportional damping, given (with the notation from above) by the sesquilinear form

$$d(t; \phi, \psi) = \int_0^l k(t, \xi) \phi(\xi) \psi(\xi) d\xi, \qquad (9.23)$$

with $k \in C^1(0, T; L_\infty(0, l))$ denoting the damping coefficient. One can take $V_D = L_2(0, l)$ here.

Time-dependent spatial hysteresis damping: This model, without time-dependence, is discussed in [DLR], and, as noted in [BSW], it has been shown to be appropriate for composite material models where graphite fibers are embedded in an epoxy matrix. The time-dependent sesquilinear form that we consider here can now be constructed with the following compact operator $K(t)$ on $L_2(0, l)$:

$$(K(t)\phi)(\xi) = \int_0^l k(t, \xi, \zeta) \phi(\zeta) d\zeta, \qquad (9.24)$$

where the nonnegative integral kernel k belongs to $C^1(0, T; L_\infty(0, l) \times L_\infty(0, l))$. Letting

$$\nu(\xi) = \int_0^l \kappa(\xi, \zeta) d\zeta \qquad (9.25)$$

denote some material property, we can define

$$d(t; \phi, \psi) = \int_0^l (\nu(\xi) - K(t))\phi'(\xi)\psi'(\xi) d\xi, \qquad (9.26)$$

with $V_D = H^1(0, l)$.

9.3   Theorems 9.1 and 9.2: Proofs

As with the first-order systems investigated earlier in Chapter 7, we will follow a standard Galerkin approximation method (see for example [BSW, LiMag, W]) with necessary, non-trivial modifications as given in [BaPed] due to the presence of the time-dependent forms. Therefore let $\{w_j\}_1^\infty$ denote a basis (i.e., a linearly independent total set) in V that is also a basis in H. This is possible since V is dense in H. For a fixed m we denote by V_m the finite dimensional subspace spanned by $\{w_j\}_1^m$, and we let y_m^0 and y_m^1 be chosen in V_m such that

$$y_m^0 \to y^0 \quad \text{in } V, \quad y_m^1 \to y^1 \quad \text{in } H, \text{ for } m \to \infty. \qquad (9.27)$$

We next define the approximate solution $y_m(t)$ of order m of our problem in the following way:

$$y_m(t) = \sum_{j=1}^m g_{jm}(t) w_j, \qquad (9.28)$$

where the $g_{jm}(t)$ are determined uniquely from the m-dimensional linear system (recall that $\langle \cdot, \cdot \rangle_{V^*,V}$ reduces to $\langle \cdot, \cdot \rangle = \langle \cdot, \cdot \rangle_H$ when the arguments are in H):

$$\langle \ddot{y}_m(t), w_j \rangle + d(t; \dot{y}_m(t), w_j) + a(t; y_m(t), w_j) = \langle f(t), w_j \rangle_{V^*,V},$$
$$j = 1, 2, ..., m;$$

with $y_m(0) = y_m^0$ and $\dot{y}_m(0) = y_m^1$. Multiplying (9.29) with $\dot{g}_{jm}(t)$ and summing over j yields

$$\langle \ddot{y}_m(t), \dot{y}_m(t) \rangle + d(t; \dot{y}_m(t), \dot{y}_m(t)) + a(t; y_m(t), \dot{y}_m(t)) = \langle f(t), \dot{y}_m(t) \rangle_{V^*,V}. \qquad (9.29)$$

Now, since

$$\frac{d}{dt}a(t; y_m(t), y_m(t)) = 2\operatorname{Re} a(t; y_m(t), \dot{y}_m(t)) + \dot{a}(t; y_m(t), y_m(t)),$$
(9.30)

we see that

$$\frac{d}{dt}\{|\dot{y}_m(t)|^2 + a(t; y_m(t), y_m(t))\} + 2\operatorname{Re} d(t; \dot{y}_m(t), \dot{y}_m(t)) =$$
$$\dot{a}(t; y_m(t), y_m(t)) + 2\operatorname{Re}\langle f(t), \dot{y}_m(t)\rangle_{V^*,V}.$$

By integrating this equality we find

$$|\dot{y}_m(t)|^2 + a(t; y_m(t), y_m(t)) + \int_0^t 2\operatorname{Re} d(t; \dot{y}_m(s), \dot{y}_m(s))ds =$$
$$|\dot{y}_m(0)|^2 + a(0; y_m^0, y_m^0) + \int_0^t \dot{a}(s; y_m(s), y_m(s))ds$$
$$+ \int_0^t 2\operatorname{Re}\langle f(s), \dot{y}_m(s)\rangle_{V^*,V}ds.$$

Using the coercivity conditions for a and d, together with the inequality (recall that $f(s) \in V_D^*$)

$$|\langle f(s), \dot{y}_m(s)\rangle_{V^*,V}| \le \frac{1}{4\epsilon}|f(s)|^2_{V_D^*} + \epsilon|\dot{y}_m(s)|^2_{V_D}$$
(9.31)

we obtain, for all $\epsilon > 0$

$$|\dot{y}_m(t)|^2 + \alpha|y_m(t)|^2_V + \int_0^t 2(\alpha_d - \epsilon)|\dot{y}_m(s)|^2_{V_D}ds \le |y_m^1|^2 + c_1|y_m^0|^2_V +$$
$$c_2 \int_0^t |y_m(s)|^2_V ds + 2\lambda_d \int_0^t |\dot{y}_m(s)|^2 ds + \int_0^t \frac{1}{2\epsilon}|f(s)|^2_{V_D^*}ds.$$

Since $y_m^0 \to y^0$ in V, $y_m^1 \to y^1$ in H and $f \in L_2(0,T;V_D^*)$, we have that, for $\epsilon > 0$ fixed and m large, there exists a constant $C > 0$, such that

$$|y_m^1|^2 + c_1|y_m^0|^2_V + \int_0^t \frac{1}{2\epsilon}|f(s)|^2_{V_D^*}ds \le C,$$
(9.32)

hence

$$|\dot{y}_m(t)|^2 + \alpha|y_m(t)|^2_V + \int_0^t 2(\alpha_d - \epsilon)|\dot{y}_m(s)|^2_{V_D}ds \le$$
$$C + c_2 \int_0^t |y_m(s)|^2_V ds + 2\lambda_d \int_0^t |\dot{y}_m(s)|^2 ds.$$

Then, in particular

$$|\dot{y}_m(t)|^2 + \alpha|y_m(t)|_V^2 \le$$
$$C + c_2 \int_0^t |y_m(s)|_V^2 ds + 2\lambda_d \int_0^t |\dot{y}_m(s)|^2 ds. \tag{9.33}$$

By Gronwall's inequality we then see that the sequence $\{\dot{y}_m\}$ is bounded in $C(0, T; H)$ and that the sequence $\{y_m\}$ is bounded in $C(0, T; V)$. From this fact together with the inequality (9.33) we conclude that $\{\dot{y}_m\}$ is also bounded in $L_2(0, T; V_D)$. Then it is possible to extract a subsequence $\{y_{m_k}\} \subset \{y_m\}$ and functions $y \in L_2(0, T; V)$ and $\tilde{y} \in L_2(0, T; V_D)$, such that $y_{m_k} \rightharpoonup y$, weakly in $L_2(0, T; V)$ and $\dot{y}_{m_k} \rightharpoonup \tilde{y}$, weakly in $L_2(0, T; V_D)$. But for $0 \le t < T$ we have in V, hence in V_D and H, that

$$y_{m_k}(t) = y_{m_k}(0) + \int_0^t \dot{y}_{m_k}(s)ds. \tag{9.34}$$

But $y_{m_k}(0) \to y^0$ in V and hence in V_D, while, for t fixed, $\int_0^t \dot{y}_{m_k}(s)ds \rightharpoonup \int_0^t \tilde{y}(s)ds$, weakly in V_D. So, by taking the weak limit in V_D in (9.34), we obtain in V_D the equality

$$y(t) = y^0 + \int_0^t \tilde{y}(s)ds, \tag{9.35}$$

from which we conclude that $\dot{y}(t)$ is in V_D a.e., with $\dot{y} = \tilde{y}$ and $y(0) = y^0$.

We need now to show that y is actually a solution to the problem (9.11), with $\dot{y}(0) = y^1$. To see this, take a function $\varphi \in C^1([0, T])$, satisfying $\varphi(T) = 0$, and define, for $j < m$, the function φ_j by $\varphi_j(t) = \varphi(t)w_j$, where $\{w_j\}_1^m$ was the basis spanning V_m. Now, for a fixed $j < m$, we multiply (9.29) with $\overline{\varphi}(t)$ and integrate to obtain

$$\int_0^T \{\langle \ddot{y}_m(s), \varphi_j(s)\rangle + d(s; \dot{y}_m(s), \varphi_j(s)) + a(s; y_m(s), \varphi_j(s))\}ds =$$
$$\int_0^T \langle f(s), \varphi_j(s)\rangle_{V_D^*, V_D} ds.$$

Observing that, for each t, we have $d(t; \cdot, \varphi_j(t)) \in V_D^*$ and $a(t; \cdot, \varphi_j(t)) \in V^*$, we find, using the weak convergence above, that for $m = m_k \to \infty$

and integration by parts in the first term, that

$$\int_0^T \{-\langle \dot{y}(s), \dot{\varphi}_j(s)\rangle + d(s; \dot{y}(s), \varphi_j(s)) + a(s; y(s), \varphi_j(s))\} \, ds =$$

$$\int_0^T \langle f(s), \varphi_j(s)\rangle_{V_D^*, V_D} \, ds + \langle y^1, \varphi_j(0)\rangle, \tag{9.36}$$

for every j. Now further restrict φ to also satisfy $\varphi \in C_0^\infty(0, T)$ and write (9.36) as

$$\int_0^T \dot{\overline{\varphi}}(s)\langle -\dot{y}(s), w_j\rangle \quad +$$

$$\int_0^T \overline{\varphi}(s)\{d(s; \dot{y}(s), w_j) + a(s; y(s), w_j) - \langle f(s), w_j\rangle_{V_D^*, V_D}\} \, ds = 0, \tag{9.37}$$

for each j. But by (9.37), we have

$$\frac{d}{dt}\langle \dot{y}(t), w_j\rangle + d(t; \dot{y}(t), w_j) + a(t; y(t), w_j) = \langle f(t), w_j\rangle_{V_D^*, V_D} \tag{9.38}$$

for all w_j. By density of $\cup_m^\infty V_m$ in V we conclude that $\ddot{y} \in L_2(0, T; V^*)$ and that for all $\psi \in V$

$$\langle \ddot{y}(t), \psi\rangle_{V^*, V} + d(t; \dot{y}(t), \psi) + a(t; y(t), \psi) = \langle f(t), \psi\rangle_{V_D^*, V_D}, \tag{9.39}$$

which was (9.11). Hence the y we have constructed is indeed a solution to the equation and by (9.35) we have that $y(0) = y^0$. In order to verify that $\dot{y}(0) = y^1$ we integrate by parts in (9.36), and by application of (9.38) we find that, for all j:

$$-\langle \dot{y}(s), \varphi_j(s)\rangle|_{s=0}^{s=T} = \langle y^1, \varphi_j(0)\rangle, \tag{9.40}$$

or, equivalently

$$\langle \dot{y}(0), w_j\rangle \overline{\varphi}(0) = \langle y^1, w_j\rangle \overline{\varphi}(0). \tag{9.41}$$

Hence $\dot{y}(0) = y^1$.

In order to prove uniqueness, let y be a solution of our problem (9.11) corresponding to $(y^0, y^1, f) = (0, 0, 0)$, and define for a fixed $t_1 \in (0, T)$ (arbitrarily chosen) the function ψ by

$$\psi(t) = \begin{cases} -\int_t^{t_1} y(s) ds & \text{for } t < t_1, \\ 0 & \text{for } t \geq t_1, \end{cases} \tag{9.42}$$

so $\psi(T) = 0$. Obviously $\psi(t) \in V$ for all t, so we can take $\psi(t) = \psi$ in (9.11) which yields

$$\langle \ddot{y}(t), \psi(t) \rangle_{V^*,V} + d(t; \dot{y}(t), \psi(t)) + a(t; y(t), \psi(t)) = \langle f(t), \psi(t) \rangle_{V^*,V}.$$
(9.43)

Because $\dot{\psi}(t) = y(t)$ for $t < t_1$ (a.e), we have that

$$\int_0^{t_1} (\langle \ddot{y}(t), \psi(t) \rangle_{V^*,V} + \langle \dot{y}(t), y(t) \rangle_{V^*,V}) dt =$$

$$\int_0^{t_1} \frac{d}{dt} (\langle \dot{y}(t), \psi(t) \rangle_{V^*,V}) dt = 0,$$
(9.44)

due to $\psi(t_1) = 0$ and the initial conditions. Using this and by integration of (9.43) we find

$$\int_0^{t_1} (\langle \dot{y}(t), y(t) \rangle_{V^*,V} - d(t; \dot{y}(t), \psi(t)) - a(t; y(t), \psi(t))) dt = 0; \quad (9.45)$$

hence

$$\int_0^{t_1} \frac{d}{dt} (|y(t)|^2 - a(t; \psi(t), \psi(t))) dt =$$

$$2 \int_0^{t_1} (\dot{a}(t; \psi(t), \psi(t)) + \operatorname{Re} d(t; \dot{y}(t), \psi(t))) dt.$$
(9.46)

Because $\psi(t_1) = 0$ and $y(0) = y^0 = 0$ this yields

$$|y(t_1)|^2 + a(0; \psi(0), \psi(0)) = 2 \int_0^{t_1} (\dot{a}(t; \psi(t), \psi(t)) + \operatorname{Re} d(t; \dot{y}(t), \psi(t))) dt.$$
(9.47)

From the assumptions on a and $\dot{a}$ we arrive at

$$|y(t_1)|^2 + \alpha|\psi(0)|_V^2 \le 2 \int_0^{t_1} (c_2|\psi(t)|_V^2 + \operatorname{Re} d(t; \dot{y}(t), \psi(t))) dt. \quad (9.48)$$

Now notice that

$$d(t; \dot{y}(t), \psi(t)) = \frac{d}{dt}(d(t; y(t), \psi(t))) - \dot{d}(t; y(t), \psi(t)) - d(t; y(t), y(t)),$$
(9.49)

so (from the initial conditions)

$$\int_0^{t_1} d(t; \dot{y}(t), \psi(t)) dt = \int_0^{t_1} (-\dot{d}(t; y(t), \psi(t)) - d(t; y(t), y(t))) dt.$$
(9.50)

Because
$$- \operatorname{Re} d(t; y(t), y(t)) \le \lambda_D |y(t)|^2 - \alpha_d |y(t)|^2_{V_D} \tag{9.51}$$

we have that

$$|y(t_1)|^2 + \alpha |\psi(0)|^2_V \le$$
$$2 \int_0^{t_1} (c_2 |\psi(t)|^2_V + \lambda_D |y(t)|^2 - \alpha_d |y(t)|^2_{V_D} + \operatorname{Re} \dot{d}(t; y(t), \psi(t))) dt. \tag{9.52}$$

Now we introduce the function $w(t) = \int_0^t y(s) ds$ and use

$$|\psi(t)|^2_V = |w(t) - w(t_1)|^2_V \le 2 |w(t)|^2_V + 2|w(t_1)|^2_V \tag{9.53}$$

to obtain

$$|y(t_1)|^2 + (\alpha - 4 c_2 t_1)|w(t_1)|^2_V \le$$
$$2 \int_0^{t_1} (2 c_2 |w(t)|^2_V + \lambda_D |y(t)|^2 - \alpha_d |y(t)|^2_{V_D} + \operatorname{Re} \dot{d}(t; y(t), \psi(t))) dt. \tag{9.54}$$

Finally we use the differentiability of the damping sesquilinear form d:

$$|\dot{d}(t; y(t), \psi(t))| \le c_4 |y(t)|_{V_D} |\psi(t)|_{V_D}$$
$$\le \frac{c_4}{2} (\epsilon |y(t)|^2_{V_D} + \frac{1}{\epsilon} |\psi(t)|^2_{V_D})$$
$$\le \frac{c_4}{2} (\epsilon |y(t)|^2_{V_D} + \frac{2}{\epsilon} |w(t)|^2_{V_D} + \frac{2}{\epsilon} |w(t_1)|^2_{V_D})$$

for all $\epsilon > 0$. We now have an inequality

$$|y(t_1)|^2 + (\alpha - 4 c_2 t_1)|w(t_1)|^2_V - \frac{2 c_4 t_1}{\epsilon} |w(t_1)|^2_{V_D} \le$$
$$2 \int_0^{t_1} (2 c_2 |w(t)|^2_V + \lambda_D |y(t)|^2 - \alpha_d |y(t)|^2_{V_D} + \frac{c_4}{2} (\epsilon |y(t)|^2_{V_D} + \frac{2}{\epsilon} |w(t)|^2_{V_D})) dt \tag{9.55}$$

and using here that

$$(\alpha - 4 c_2 t_1)|w(t_1)|^2_V \ge (\frac{\alpha}{2} - 2 c_2 t_1)(|w(t_1)|^2_V + |w(t_1)|^2_{V_D}), \tag{9.56}$$

we finally arrive at the formidable inequality

$$|y(t_1)|^2 + (\frac{\alpha}{2} - 2c_2t_1)|\omega(t_1)|_V^2 + (\frac{\alpha}{2} - 2c_2t_1 - \frac{2c_4t_1}{\epsilon})|\omega(t_1)|_{V_D}^2 \le$$
$$2\int_0^{t_1}(2c_2|\omega(t)|_V^2 + \lambda_D|y(t)|^2 + (\frac{c_4\epsilon}{2} - \alpha_d)|y(t)|_{V_D}^2 + \frac{c_4}{\epsilon}|\omega(t)|_{V_D}^2)dt.$$

Now fix $\epsilon > 0$ such that $(\frac{c_4\epsilon}{2} - \alpha_d) < 0$ and fix $t_1 = \frac{\alpha}{8(c_2+\frac{c_4}{\epsilon})}$ such that
$(\frac{\alpha}{2} - 2c_2t_1 - \frac{2c_4t_1}{\epsilon}) = \frac{\alpha}{4}$. Then also $(\frac{\alpha}{2} - 2c_2t_1) > 0$ and the inequality
above implies that for some constant $M > 0$,

$$|y(t_1)|^2 + |\omega(t_1)|_V^2 + |\omega(t_1)|_{V_D}^2 \le M\int_0^{t_1}(|y(t)|^2 + |\omega(t)|_V^2 + |\omega(t)|_{V_D}^2)dt.$$

$$(9.57)$$

From Gronwall's inequality we see that $y = 0$ in the interval $[0, t_1]$.
Since the length of t_1 is independent of the choice of origin, we conclude that $y = 0$ on $[t_1, 2t_1]$, etc. Hence $y = 0$ and uniqueness is
proved. That the solution depends continuously on the data is obvious
from the inequalities used to show existence; indeed, from (9.32) and
(9.33) and the weak lower semicontinuity of norms we conclude that
the constructed solution satisfies

$$|y(t)|_V^2 + |\dot{y}(t)|^2 + \delta\int_0^t|\dot{y}(t)|_{V_D}^2 \le$$
$$K\left(|y^0|_V^2 + |y^1|^2 + \int_0^t|f(s)|_{V_D^*}^2 ds\right) \quad (9.58)$$

for some positive constants δ and K. Integrating from 0 to T yields
the desired result (since $(y^0, y^1, f) \to (y, \dot{y})$ is linear). This completes
the proof of Theorem 9.1.

The proof of Theorem 9.2 follows from the inequality (9.57) and the
original proof in the case $d = 0$ from [LiMag, p. 275–279], because we
do not gain any additional regularity from the form of d in this case.

10   Inverse or Parameter Estimation Problems

In the generic abstract parameter estimation problem, we consider a dynamic model of the form (9.1) where the operators A and D and possibly the input f depend on some unknown (i.e., to be estimated) functional parameters q in an admissible family $\mathcal{Q} \subset C^1(0,T;L_\infty(\Omega;Q))$ of parameters. Here Ω is the underlying set on which the functions of H, V, V_D are defined (e.g., the spatial set $\Omega = (0,l)$ in the heat, transport, and beam examples). We assume that the time dependence of the operators A and D are through the time dependence of the parameters $q(t) \in L_\infty(\Omega;Q)$ where $Q \subset R^p$ is a given constraint set for the values of the parameters. That is, $A(t) = A_1(q(t)), D(t) = A_2(q(t))$ so that we have

$$\ddot{y}(t) + A_2(q(t))\dot{y}(t) + A_1(q(t))y(t) = f(t,q) \quad \text{in } V^*$$
$$y(0) = y^0, \quad \dot{y}(0) = y^1. \tag{10.1}$$

Thus we introduce the sesquilinear forms σ_1, σ_2 by

$$a(t;\phi,\psi) \equiv \sigma_1(q(t))(\phi,\psi) = \langle A_1(q(t))\phi,\psi \rangle_{V^*,V}, \tag{10.2}$$
$$d(t;\phi,\psi) \equiv \sigma_2(q(t))(\phi,\psi) = \langle A_2(q(t))\phi,\psi \rangle_{V_D^*,V_D}. \tag{10.3}$$

It will be convenient in subsequent arguments to use the notation

$$\dot{\sigma}_i(q)(\phi,\psi) \equiv \frac{d}{dt}\sigma_i(q)(\phi,\psi) \tag{10.4}$$

which in the event that σ_i is linear in q becomes $\dot{\sigma}_i(q)(\phi,\psi) = \sigma_i(\dot{q})(\phi,\psi)$. For example, in the Euler-Bernoulli example of (9.18), we would have

$$\sigma_1(q(t))(\phi,\psi) = \int_0^l \widetilde{EI}(t,\xi)\phi''(\xi)\psi''(\xi)d\xi \tag{10.5}$$

$$\sigma_2(q(t))(\phi,\psi) = \int_0^l \widetilde{C_DI}(t,\xi)\phi''(\xi)\psi''(\xi)d\xi, \tag{10.6}$$

where $q = (\widetilde{EI}, \widetilde{C_DI}) \in C^1(0,T;L^\infty(0,l;R_+^2))$. Note that in this case we do have $\dot{\sigma}_i(q)(\phi,\psi) = \sigma_i(\dot{q})(\phi,\psi)$.

In terms of parameter dependent sesquilinear forms we thus will write (10.1) as

$$\langle \ddot{y}(t),\phi \rangle + \sigma_2(q(t))(\dot{y}(t),\phi) + \sigma_1(q(t))(y(t),\phi) = \langle f(t,q),\phi \rangle$$
$$y(0) = y^0, \quad \dot{y}(0) = y^1 \tag{10.7}$$

for all $\phi \in V$. As in (9.2), $\langle \cdot, \cdot \rangle$ denotes the duality product $\langle \cdot, \cdot \rangle_{V^*, V}$. In some problems the initial data y^0, y^1 may also depend on parameters $\tilde{q}$ to be estimated, i.e., $y^0 = y^0(\tilde{q}), y^1 = y^1(\tilde{q})$. We shall not discuss such problems here, although the ideas we present can be used to effectively treat such problems. We instead refer readers to [BK] for discussions of general estimation problems where not only the initial data but even the underlying spaces V and H themselves may depend on unknown parameters.

It is assumed that the parameter-dependent sesquilinear forms $\sigma_1(q), \sigma_2(q)$ of (10.7) satisfy the continuity and ellipticity conditions (H2)-(H7) of Section 9.1 uniformly in $q \in \mathcal{Q}$; that is, the constants $c_1, c_2, \alpha, c_3, c_4, \lambda_d, \alpha_d$ of (H2)-(H7) can be found independently of $q \in \mathcal{Q}$.

In general inverse problems, one must estimate the functional parameters q from dynamic observations of the system (10.1) or (10.7). A fundamental consideration in problem formulation involves what will be measured in the dynamic experiments producing the observations. To discuss these measurements in a specific setting, we consider the transverse vibrations (for example, equation (9.18)) of our beam example where $y(t) = y(t, \cdot)$. Measurements, of course, depend on the sensors available. If one considers a truly smart material structure as in [BSW, RCS], it contains both sensors and actuators which may or may not rely on the same physical device or material. In usual mechanical experiments, there are several popular measurement devices [BSW, BT], some of which could possibly be used in a smart material configuration. If one uses an accelerometer placed at the point $\bar{\xi} \in (0, \ell)$ along the beam, then one obtains observations $\ddot{y}(t, \bar{\xi})$ of beam *acceleration*. A laser vibrometer will yield data of *velocity* $\dot{y}(t, \bar{\xi})$ while proximity probes including displacement solenoids produce measurements of *displacement* $y(t, \bar{\xi})$. In the case of a beam (or structure) with piezoceramic patches, the patches may be used as sensors as well as actuators. In this case one obtains observations of voltages which are proportional to the *accumulated strain*; this is discussed fully in [BSW, RCS].

Whatever the measuring devices, the resulting observations can be used in a maximum likelihood or some type of least squares formulation of the parameter estimation problem, depending on assumptions about the statistical model [BDSS, BT] for errors in the observation process. In the least squares formulations, the problems are stated in terms of finding parameters which minimize the sum-of-squares distance between the parameter-dependent solutions of the partial differential equation and dynamic system response data collected after various excitations, while the maximum likelihood estimator results from maximizing a given (assumed) likelihood function for the parameters, given the data.

In the beam example, the parameters to be estimated include the stiffness coefficient $\widetilde{EI}(t, \xi)$, the Kelvin-Voigt damping parameter $\widetilde{c_D I}(t, \xi)$, and any control related parameters that arise in the actuator input f. Details regarding the estimation of these parameters in the time independent case for an experimental beam are given in Section 5.4 of [BSW], while similar experimental results for a plate are summarized in Section 5.5 of the same reference.

The general ordinary least squares parameter estimation problem can be formulated as follows. For a given discrete set of measured observations $z = \{z_i\}_{i=1}^{N_t}$ corresponding to model observations $z_{ob}(t_i)$ at times t_i as obtained in most practical cases, we consider the problem of minimizing over $q \in Q$ the least squares output functional

$$J(q, z) = \left| \tilde{C}_2 \left\{ \tilde{C}_1 \{ y(t_i, \cdot; q) \} - \{ z_i \} \right\} \right|^2, \qquad (10.8)$$

where $\{y(t_i, \cdot; q)\}$ are the parameter dependent solutions of (10.1) or (10.7) evaluated at each time $t_i, i = 1, 2, \ldots, N_t$ and $| \cdot |$ is an appropriately chosen Euclidean norm. Here the operators $\tilde{C}_1$ and $\tilde{C}_2$ are observation operators that depend on the type of observed or measured data available. The operator $\tilde{C}_1$ may have several forms depending on the type of sensors being used. When the collected data z_i consists of time domain displacement, velocity, or acceleration values at a point $\bar{\xi}$ on the beam as discussed above, the functional takes the form

$$J_\nu(q, z) = \sum_{i=1}^{N_t} \left| \frac{\partial^\nu y}{\partial t^\nu}(t_i, \bar{\xi}; q) - z_i \right|^2, \qquad (10.9)$$

for $\nu = 0, 1, 2$, respectively. In this case the operator $\tilde{C}_1$ involves differentiation (either $\nu = 0, 1$ or 2 times, respectively) with respect to time followed by pointwise evaluation in t and $\bar{\xi}$. We shall in our presentation of the next section adopt the ordinary least squares functional (10.8) to formulate and develop our results.

10.1   Approximation and Convergence

In this section we present a *corrected version* (Theorem 5.2 of [BSW] contains errors in statement and proof) and extension (to treat time dependent coefficients) of arguments for approximation and convergence in inverse problems found in Section 5.2 of [BSW]. For more details on general inverse problem methodology in the context of abstract structural systems, the reader may consult [BSW]. The Banks-Kunisch book [BK] contains a general treatment of inverse problems for partial differential equations in a functional analytic setting.

The minimization in our general abstract parameter estimation problems for (10.8) involves an infinite dimensional state space H and an infinite dimensional admissible parameter set Q (of functions). To obtain computationally tractable methods, we thus consider Galerkin type approximations in the context of the variational formulation (10.7). Let H^N be a sequence of finite dimensional subspaces of H, and Q^M be a sequence of finite dimensional sets approximating the parameter set Q. We denote by P^N the orthogonal projections of H onto H^N. Then a family of approximating estimation problems with finite dimensional state spaces and parameter sets can be formulated by seeking $q \in Q^M$ which minimizes

$$J^N(q, z) = \left| \tilde{C}_2 \left\{ \tilde{C}_1 \{ y^N(t_i, \cdot; q) \} - \{z_i\} \right\} \right|^2, \qquad (10.10)$$

where $y^N(t; q) \in H^N$ is the solution to the finite dimensional approximation of (10.7) given by

$$\langle \ddot{y}^N(t), \phi \rangle + \sigma_2(q(t))(\dot{y}^N(t), \phi) + \sigma_1(q(t))(y^N(t), \phi) = \langle f(t, q), \phi \rangle$$
$$y^N(0) = P^N y^0, \quad \dot{y}^N(0) = P^N y^1,$$

$$(10.11)$$

for $\phi \in H^N$. For the parameter sets Q and Q^M, and state spaces H^N, we make the following hypotheses which detail approximation conditions with respect to the M (parameter set) and N (state space) indices.

(A1M) The sets Q and Q^M lie in a metric space $\tilde{Q}$ with metric d. It is assumed that Q and Q^M are compact in this metric and there is a mapping $i^M : Q \to Q^M$ so that $Q^M = i^M(Q)$. Furthermore, for each $q \in Q$, $i^M(q) \to q$ in $\tilde{Q}$ with the convergence uniform in $q \in Q$.

(A2N) The finite dimensional subspaces H^N satisfy $H^N \subset V$ as well as the approximation properties of the next two statements.

(A3N) For each $\psi \in V, |\psi - P^N \psi|_V \to 0$ as $N \to \infty$.

(A4N) For each $\psi \in V_D, |\psi - P^N \psi|_{V_D} \to 0$ as $N \to \infty$.

The reader is referred to Chapter 4 of [BSW] for a complete discussion motivating the spaces H^N and V_D.

We also need some regularity with respect to the parameters q in the parameter dependent sesquilinear forms σ_1, σ_2. In addition to (uniform in $\mathcal{Q}$) ellipticity/coercivity and continuity conditions (H2)-(H7), the sesquilinear forms $\sigma_1 = \sigma_1(q)$, $\sigma_2 = \sigma_2(q)$ and $\dot{\sigma}_1 = \dot{\sigma}_1(q)$ are assumed to be defined on $\mathcal{Q}$ and satisfy the continuity-with-respect-to-parameter conditions

(H8) $|\sigma_1(q)(\phi, \psi) - \sigma_1(\tilde{q})(\phi, \psi)| \le \gamma_1 d(q, \tilde{q})|\phi|_V |\psi|_V$, for $\phi, \psi \in V$

(H9) $|\dot{\sigma}_1(q)(\phi, \psi) - \dot{\sigma}_1(\tilde{q})(\phi, \psi)| \le \gamma_3 d(q, \tilde{q})|\phi|_V |\psi|_V$, for $\phi, \psi \in V$

(H10) $|\sigma_2(q)(\xi, \eta) - \sigma_2(\tilde{q})(\xi, \eta)| \le \gamma_2 d(q, \tilde{q})|\xi|_{V_D} |\eta|_{V_D}$, for $\xi, \eta \in V_D$

for $q, \tilde{q} \in \mathcal{Q}$ where the constants $\gamma_1, \gamma_2, \gamma_3$ depend only on $\mathcal{Q}$.

Solving the approximate estimation problems involving (10.10),(10.11), we obtain a sequence of parameter estimates $\{\bar{q}^{N,M}\}$. It is of paramount importance to establish conditions under which $\{\bar{q}^{N,M}\}$ (or some subsequence) converges to a solution for the original infinite dimensional estimation problem involving (10.7),(10.8). Toward this goal we have the following results.

Theorem 10.1 *To obtain convergence of at least a subsequence of* $\{\bar{q}^{N,M}\}$ *to a solution* $\bar{q}$ *for the problem of minimizing* (10.8) *subject to* (10.7), *it suffices, under assumption (A1M), to argue that for arbitrary sequences* $\{q^{N,M}\}$ *in* $\mathcal{Q}^M$ *with* $q^{N,M} \to q$ *in* $\mathcal{Q}$, *we have*

$$\tilde{C}_2 \tilde{C}_1 y^N(t; q^{N,M}) \to \tilde{C}_2 \tilde{C}_1 y(t; q). \tag{10.12}$$

Proof Under the assumptions (A1M), let $\{\bar{q}^{N,M}\}$ be solutions minimizing (10.10) subject to the finite dimensional system (10.11) and let $\hat{q}^{N,M} \in \mathcal{Q}$ be such that $i^M(\hat{q}^{N,M}) = \bar{q}^{N,M}$. From the compactness of $\mathcal{Q}$, we may select subsequences, again denoted by $\{\hat{q}^{N,M}\}$ and $\{\bar{q}^{N,M}\}$, so that $\hat{q}^{N,M} \to \bar{q} \in \mathcal{Q}$ and $\bar{q}^{N,M} \to \bar{q}$ (the latter follows the last statement of (A1M)). The optimality of $\{\bar{q}^{N,M}\}$ guarantees that for every $q \in \mathcal{Q}$

$$J^N(\bar{q}^{N,M}, z) \le J^N(i^M(q), z). \tag{10.13}$$

Using (10.12), the last statement of (A1M) and taking the limit as $N, M \to \infty$ in the inequality (10.13), we obtain $J(\bar{q}, z) \le J(q, z)$ for every $q \in \mathcal{Q}$, or that $\bar{q}$ is a solution of the problem for (10.7),(10.8). We note that under uniqueness assumptions on the problems (a situation that we hasten to add is not often realized in practice), one can actually guarantee convergence of the entire sequence $\{\bar{q}^{N,M}\}$ in place of subsequential convergence to solutions.

We note that the essential aspects in the arguments given above involve compactness assumptions on the sets $\mathcal{Q}^M$ and $\mathcal{Q}$. Such compactness ideas play a fundamental role in other theoretical and computational aspects of these problems. For example, one can formulate distinct concepts of *problem stability* and *method stability* as in [BK] involving some type of continuous dependence of solutions on the observations z, and use conditions similar to those of (10.12) and (A1M), with compactness again playing a critical role, to guarantee stability. We illustrate with a simple form of *method stability* (other stronger forms are also amenable to this approach–see [BK]).

We might say that an *approximation method*, such as that formulated above involving $\mathcal{Q}^M, H^N$ and (10.10), is *stable* if

$$dist(\tilde{q}^{N,M}(z^k), \tilde{q}(z^*)) \to 0$$

as $N, M, k \to \infty$ for any $z^k \to z^*$ (in this case in the appropriate Euclidean space), where $\tilde{q}(z)$ denotes the set of all solutions of the problem for (10.8) and $\tilde{q}^{N,M}(z)$ denotes the set of all solutions of the problem for (10.10). Here "dist" represents the usual set distance function. Under (10.12) and (A1M), one can use arguments very similar to those sketched above to establish that one has this method stability. If the sets $\mathcal{Q}^M$ are not defined through a mapping i^M as supposed above, one can still obtain this method stability if one replaces the last statement of (A1M) by the assumptions:

(i) If $\{q^M\}$ is *any* sequence with $q^M \in \mathcal{Q}^M$, then there exist q^* in $\mathcal{Q}$ and subsequence $\{q^{M_k}\}$ with $q^{M_k} \to q^*$ in the $\tilde{\mathcal{Q}}$ topology.

(ii) For *any* $q \in \mathcal{Q}$, there exists a sequence $\{q^M\}$ with $q^M \in \mathcal{Q}^M$ such that $q^M \to q$ in $\tilde{\mathcal{Q}}$.

Similar ideas may be employed to discuss the question of *problem stability* for the problem of minimizing (10.8) over $\mathcal{Q}$ (i.e., the original problem) and again compactness of the admissible parameter set plays a critical role.

Compactness of parameter sets also plays an important role in computational considerations. In certain problems, the formulation outlined above (involving $\mathcal{Q}^M = i^M(\mathcal{Q})$) results in a computational framework wherein the $\mathcal{Q}^M$ and $\mathcal{Q}$ all lie in some uniform set possessing compactness properties. The compactness criteria can then be reduced to uniform constraints on the derivatives of the admissible parameter functions. There are numerical examples (for example, see [BI86]) which demonstrate that imposition of these constraints is necessary (and sufficient) for convergence of the resulting algorithms. (This offers a possible explanation for some of the numerical failures [YY] of such methods reported in the engineering literature.)

Thus we have that compactness of admissible parameter sets plays a fundamental role in a number of aspects, both theoretical and computational, in parameter estimation problems. This compactness may be assumed (and imposed) explicitly as we have outlined here, or it may be included implicitly in the problem formulation through *Tikhonov regularization* as discussed for example by Kravaris and Seinfeld [KS], Vogel [Vog], and widely by many others. In the regularization approach one restricts consideration to a subset $\mathcal{Q}_1$ of parameters which has compact embedding in $\mathcal{Q}$ and modifies the least-squares criterion to include a term which insures that minimizing sequences will be $\mathcal{Q}_1$ bounded and hence compact in the original parameter set $\mathcal{Q}$.

After this short digression on general inverse problem concepts, we return to the condition (10.12). To demonstrate that this condition can be readily established in many problems of interest to us here, we give the following general convergence results. We remark that this is a restated and corrected version of Theorem 5.2 of [BSW].

Theorem 10.2 *Suppose that H^N satisfies (A2N),(A3N),(A4N) and assume that the sesquilinear forms $\sigma_1(q), \dot{\sigma}_1(q)$ and $\sigma_2(q)$ satisfy (H8),(H9),(H10), respectively, as well as (H1)-(H7) of Section 9.1 (uniformly in $q \in \mathcal{Q}$). Furthermore, assume that*

$$q \to f(\cdot; q) \text{ is continuous from } \mathcal{Q} \text{ to } L_2(0, T; V_D^*). \qquad (10.14)$$

Let q^N be arbitrary in $\mathcal{Q}$ such that $q^N \to q$ in $\mathcal{Q}$. Then if in addition

$\dot{y} \in L_2(0, T; V)$, *we have as* $N \to \infty$,

$$y^N(t; q^N) \to y(t; q) \quad \text{in } V \text{ norm for each } t > 0$$
$$\dot{y}^N(t; q^N) \to \dot{y}(t; q) \quad \text{in } L_2(0, T; V_D) \cap C(0, T; H),$$

where $(y^N, \dot{y}^N)$ *are the solutions to* (10.11) *and* $(y, \dot{y})$ *are the solutions to* (10.7).

Proof From Theorem 9.2 above we find that the solution of (10.7) satisfies $y(t) \in V$ for each t, $\dot{y}(t) \in V_D$ for almost every $t > 0$. Because

$$|y^N(t; q^N) - y(t; q)|_V \le |y^N(t; q^N) - P^N y(t; q)|_V + |P^N y(t; q) - y(t; q)|_V,$$

and (A3N) implies (recall that P^N is the orthogonal projection of H onto H^N of (A2N)-(A4N)) that the second term on the right side converges to 0 as $N \to \infty$, it suffices for the first convergence statement to show that

$$|y^N(t; q^N) - P^N y(t; q)|_V \to 0 \quad \text{as } N \to \infty.$$

Similarly, we note that this same inequality with y^N, y replaced by $\dot{y}^N, \dot{y}$ and the V-norm replaced by the V_D-norm along with (A4N) permits us to claim that the convergence

$$|\dot{y}^N(t; q^N) - P^N \dot{y}(t; q)|_{V_D} \to 0 \quad \text{as } N \to \infty,$$

is sufficient to establish the second convergence statement of the theorem. We shall, in fact, establish the convergence of $\dot{y}^N - P^N \dot{y}$ in the stronger V norm.

Let $y^N = y^N(t; q^N)$, $y = y(t; q)$, and $\Delta^N = \Delta^N(t) \equiv y^N(t; q^N) - P^N y(t; q)$. Then

$$\dot{\Delta}^N = \dot{y}^N - \frac{d}{dt} P^N y = \dot{y}^N - P^N \dot{y}$$

and

$$\ddot{\Delta}^N = \ddot{y}^N - \frac{d^2}{dt^2} P^N y$$

because $\dot{y} \in L_2((0, T), V_D)$, $\ddot{y} \in L_2((0, T), V^*)$. We suppress the dependence on t in the arguments below when no confusion will result. From

(10.7) and (10.11), we have for $\psi \in H^N$

$$
\begin{aligned}
\langle \ddot{\Delta}^N, \psi \rangle_{V^*,V} \;=\; & \langle \ddot{y}^N - \ddot{y} + \ddot{y} - \frac{d^2}{dt^2} P^N y, \psi \rangle_{V^*,V} \\
=\; & \langle f(q^N), \psi \rangle_{V_D^*,V_D} - \sigma_2(q^N)(\dot{y}^N, \psi) - \sigma_1(q^N)(y^N, \psi) \\
& - \langle f(q), \psi \rangle_{V_D^*,V_D} + \sigma_2(q)(\dot{y}, \psi) + \sigma_1(q)(y, \psi) \\
& + \langle \ddot{y} - \frac{d^2}{dt^2} P^N y, \psi \rangle_{V^*,V}.
\end{aligned}
$$

This can be written as

$$
\begin{aligned}
& \langle \ddot{\Delta}^N, \psi \rangle_{V^*,V} + \sigma_1(q^N)(\Delta^N, \psi) \\
=\; & \langle \ddot{y} - \frac{d^2}{dt^2} P^N y, \psi \rangle_{V^*,V} - \langle f(q) - f(q^N), \psi \rangle_{V_D^*,V_D} \\
& + \sigma_2(q^N)(\dot{y} - P^N \dot{y}, \psi) + \sigma_2(q)(\dot{y}, \psi) - \sigma_2(q^N)(\dot{y}, \psi) \\
& + \sigma_1(q^N)(y - P^N y, \psi) + \sigma_1(q)(y, \psi) - \sigma_1(q^N)(y, \psi) \\
& - \sigma_2(q^N)(\dot{\Delta}^N, \psi).
\end{aligned} \qquad (10.15)
$$

Choosing $\dot{\Delta}^N$ as the test function ψ in (10.15) and employing the equality $\langle \ddot{\Delta}^N, \dot{\Delta}^N \rangle_{V^*,V} = \frac{1}{2} \frac{d}{dt} |\dot{\Delta}^N|_H^2$ (this follows using definitions of the duality mapping – see [BKW1] and the hypothesis (A2N)), and using the symmetry of σ_1, we have

$$
\begin{aligned}
\frac{1}{2} \frac{d}{dt} & \Big\{ |\dot{\Delta}^N|_H^2 + \sigma_1(q^N)(\Delta^N, \Delta^N) \Big\} \\
=\; & \mathrm{Re} \Big\{ \langle \ddot{y} - \frac{d^2}{dt^2} P^N y, \dot{\Delta}^N \rangle_{V^*,V} - \langle f(q) - f(q^N), \dot{\Delta}^N \rangle_{V_D^*,V_D} \\
& + \sigma_2(q^N)(\dot{y} - P^N \dot{y}, \dot{\Delta}^N) + \sigma_2(q)(\dot{y}, \dot{\Delta}^N) \\
& - \sigma_2(q^N)(\dot{y}, \dot{\Delta}^N) + \sigma_1(q^N)(y - P^N y, \dot{\Delta}^N) \\
& + \sigma_1(q)(y, \dot{\Delta}^N) - \sigma_1(q^N)(y, \dot{\Delta}^N) - \sigma_2(q^N)(\dot{\Delta}^N, \dot{\Delta}^N) \\
& \qquad\qquad + \dot{\sigma}_1(q^N)(\Delta^N, \Delta^N) \Big\}.
\end{aligned}
$$

$$(10.16)$$

We observe that $\langle \ddot{y} - \frac{d^2}{dt^2} P^N y, \dot{\Delta}^N \rangle_{V^*,V} \equiv 0$ because P^N is an orthogonal projection. Thus, we find

$$\frac{1}{2}\frac{d}{dt}\left\{|\dot{\Delta}^N|_H^2 + \sigma_1(q^N)(\Delta^N, \Delta^N)\right\}$$

$$= \text{Re}\left\{-T_3^N + \sigma_2(q^N)(\dot{y} - P^N\dot{y}, \dot{\Delta}^N) + T_2^N + \frac{d}{dt}(\sigma_1(q^N)(y - P^N y, \Delta^N))\right.$$

$$- \sigma_1(q^N)(\frac{d}{dt}(y - P^N y), \Delta^N) - \dot{\sigma}_1(q^N)(y - P^N y, \Delta^N)$$

$$\left. + T_1^N + \sigma_2(q^N)(\dot{\Delta}^N, \dot{\Delta}^N) + \dot{\sigma}_1(q^N)(\Delta^N, \Delta^N)\right\},$$

$$(10.17)$$

where

$$\begin{aligned} T_1^N &= \Delta\sigma_1^N(y, \dot{\Delta}^N) \equiv \sigma_1(q)(y, \dot{\Delta}^N) - \sigma_1(q^N)(y, \dot{\Delta}^N) \\ T_2^N &= \Delta\sigma_2^N(\dot{y}, \dot{\Delta}^N) \equiv \sigma_2(q)(\dot{y}, \dot{\Delta}^N) - \sigma_2(q^N)(\dot{y}, \dot{\Delta}^N) \qquad (10.18) \\ T_3^N &= \langle \Delta f^N, \dot{\Delta}^N \rangle_{V_D^*, V_D} \equiv \langle f(q) - f(q^N), \dot{\Delta}^N \rangle_{V_D^*, V_D}. \end{aligned}$$

Note that here we have used that $\dot{y} \in L_2(0, T; V)$. Integrating the terms in (10.17) from 0 to t and using the initial conditions

$$\Delta^N(0) = y^N(0) - P^N y(0) = y^N(0) - P^N y^0 = 0$$
$$\dot{\Delta}^N(0) = \dot{y}^N(0) - P^N \dot{y}(0) = \dot{y}^N(0) - P^N y^1 = 0,$$

we obtain (here we do include arguments (t) and (s) in our calculations and estimates to avoid confusion)

$$\frac{1}{2}|\dot{\Delta}^N(t)|_H^2 + \sigma_1(q^N(t))(\Delta^N(t), \Delta^N(t))$$

$$= \int_0^t \left\{\text{Re}\{\sigma_2(q^N(s))(\dot{y}(s) - P^N\dot{y}(s), \dot{\Delta}^N(s))\right.$$

$$- \sigma_1(q^N(s))(\dot{y}(s) - P^N\dot{y}(s), \Delta^N(s)) - \dot{\sigma}_1(q^N(s))(y(s) - P^N y(s), \Delta^N(s))$$

$$- \sigma_2(q^N(s))(\dot{\Delta}^N(s), \dot{\Delta}^N(s)) + T_1^N(s)$$

$$+ T_2^N(s) + T_3^N(s) + \dot{\sigma}_1(q^N(s))(\Delta^N(s), \Delta^N(s))\}\bigg\}ds$$

$$+ \text{Re}\left\{\sigma_1(q(t))(y(t) - P^N y(t), \Delta^N(t))\right\}. \quad (10.19)$$

We consider the T_i^N terms in this equation. We have

$$T_1^N = \frac{d}{dt}\Delta\sigma_1^N(y, \Delta^N) - \Delta\sigma_1^N(\dot{y}, \Delta^N) - \Delta\dot{\sigma}_1^N(y, \Delta^N)$$

so that

$$\int_0^t T_1^N(s)ds = \Delta\sigma_1^N(y(t),\Delta^N(t))$$

$$- \int_0^t \left\{ \Delta\sigma_1^N(\dot{y}(s),\Delta^N(s)) - \Delta\dot{\sigma}_1^N(y(s),\Delta^N(s)) \right\}ds.$$

Using (H8),(H9) we thus obtain

$$\text{Re}\int_0^t T_1^N(s)ds \le \frac{\gamma_1^2}{4\epsilon}d(q^N,q)^2|y(t)|_V^2 + \epsilon|\Delta^N(t)|_V^2$$

$$+ \int_0^t \left\{ \frac{\gamma_1^2}{4\epsilon}d(q^N,q)^2|\dot{y}(s)|_V^2 + \epsilon|\Delta^N(s)|_V^2 \right.$$

$$\left. + \frac{\gamma_3^2}{4\epsilon}d(q^N,q)^2|y(s)|_V^2 + \epsilon|\Delta^N(s)|_V^2 \right\}ds. \quad (10.20)$$

Similarly, using (H10) we find

$$\text{Re}\int_0^t (T_2^N(s) + T_3^N(s))ds \le$$

$$\int_0^t \left\{ \frac{\gamma_2^2}{4\epsilon}d(q^N,q)^2|\dot{y}(s)|_V^2 + 2\epsilon|\dot{\Delta}^N(s)|_{V_D}^2 + \frac{1}{4\epsilon}|\Delta f^N(s)|_V^2 \right\}ds. \quad (10.21)$$

These can then be used in (10.19) to obtain the estimate

$$\frac{1}{2}|\dot{\Delta}^N(t)|_H^2 + \alpha|\Delta^N(t)|_V^2 + \int_0^t \alpha_d|\dot{\Delta}^N(s)|_{V_D}^2 ds$$

$$\le \int_0^t \left\{ \lambda_d|\dot{\Delta}^N(s)|_H^2 + \frac{c_3^2}{4\epsilon}|\dot{y}(s) - P^N\dot{y}(s)|_{V_D}^2 + 3\epsilon|\dot{\Delta}^N(s)|_{V_D}^2 \right.$$

$$+ \frac{c_1^2}{2}|\dot{y}(s) - P^N\dot{y}(s)|_V^2 + \frac{c_2^2}{2}|y(s) - P^N y(s)|_V^2 + (1+\epsilon+c_2)|\Delta^N(s)|_V^2$$

$$+ \frac{\gamma_1^2}{4\epsilon}d(q^N,q)^2|\dot{y}(s)|_V^2 + \frac{\gamma_3^2}{4\epsilon}d(q^N,q)^2|y(s)|_V^2 + \frac{\gamma_2^2}{4\epsilon}d(q^N,q)^2|\dot{y}(s)|_V^2$$

$$+ \frac{1}{4\epsilon}|\Delta f^N(s)|_V^2 \bigg\}ds + \frac{c_1^2}{4\epsilon}|y(t) - P^N y(t)|_V^2 + 2\epsilon|\Delta^N(t)|_V^2 + \frac{\gamma_1^2}{4\epsilon}d(q^N,q)^2|y(t)|_V^2.$$

This finally reduces to

$$\frac{1}{2}|\dot{\Delta}^N(t)|_H^2 + (\alpha - 2\epsilon)|\Delta^N(t)|_V^2 + \int_0^t (\alpha_d - 3\epsilon)|\dot{\Delta}^N(s)|_{V_D}^2 ds$$

$$\leq \int_0^t \left\{ \lambda_d |\dot{\Delta}^N(s)|_H^2 + (1 + \epsilon + c_2)|\Delta^N(s)|_V^2 \right\} ds$$

$$+ \underbrace{ \frac{c_1^2}{4\epsilon}|y(t) - P^N y(t)|_V^2 + \int_0^t \left\{ \frac{c_3^2}{4\epsilon}|\dot{y}(s) - P^N \dot{y}(s)|_{V_D}^2 + \frac{c_1^2}{2}|\dot{y}(s) - P^N \dot{y}(s)|_V^2 \right\} }_{\delta_1^N(t)}$$

$$+ \underbrace{ \frac{\gamma_1^2}{4\epsilon}d(q^N,q)^2|y(t)|_V^2 + \int_0^t \left\{ \frac{1}{4\epsilon}|\Delta f^N(s)|_V^2 + \frac{c_2^2}{2}|y(s) - P^N y(s)|_V^2 \right\} ds }_{\delta_2^N(t)}$$

$$+ \underbrace{ \int_0^t \left\{ \frac{\gamma_1^2}{4\epsilon}d(q^N,q)^2|\dot{y}(s)|_V^2 + \frac{\gamma_3^2}{4\epsilon}d(q^N,q)^2|y(s)|_V^2 + \frac{\gamma_2^2}{4\epsilon}d(q^N,q)^2|\dot{y}(s)|_V^2 + \right\} d }_{\delta_3^N(t)}$$

Therefore, under the assumptions we have as $N \to \infty$ and $q^N \to q$

$$\sup_{t \in (0,T)} [\delta_1^N(t) + \delta_2^N(t) + \delta_3^N(t)] \to 0.$$

Applying Gronwall's inequality, we find that

$$\dot{\Delta}^N \to 0 \text{ in } C(0,T;H)$$
$$\Delta^N \to 0 \text{ in } C(0,T;V)$$
$$\dot{\Delta}^N \to 0 \text{ in } L_2(0,T;V_D),$$

and hence the convergence statement of the theorem holds.

Remark: The condition "$\dot{y} \in L_2(0,T;V)$" routinely occurs if the structure under investigation has sufficiently strong damping so that V_D is equivalent to V.

10.2   Some Further Remarks

We have provided sufficient conditions and detailed arguments for existence and uniqueness of solutions to abstract second-order non-autonomous hyperbolic systems such as those with time dependent "stiffness," "damping," and input parameters. In addition, we have considered a class of corresponding inverse problems for these systems and argued convergence results for approximating problems that yield a type of method stability as well as a framework for finite element computational techniques. The efficacy of such methods has been demonstrated for autonomous systems in earlier efforts; see [BSW] and the references therein. The approaches developed here can readily be extended (albeit with considerable tedium) for higher dimensional spatial systems such as plates, shells, etc.

11 "Weak" or "Variational Form"

We consider the origin of the terms "<u>weak</u> or <u>variational</u> form" of Chapters 7 and 9 as opposed to strong or closed form of PDE's. We use the beam equation to illustrate ideas.

Recall Example 6, the cantilever beam. This example, given in classical form (which can be derived in a straightforward manner using force and moment balance–see [BT]) is

$$\rho\frac{\partial^2 y}{\partial t^2} + \gamma\frac{\partial y}{\partial t} + \frac{\partial^2}{\partial \xi^2}\left(EI\frac{\partial^2 y}{\partial \xi^2} + c_D I\frac{\partial^3 y}{\partial \xi^2 \partial t}\right) = f(t,\xi) \qquad (11.1)$$

with boundary conditions

$$\begin{aligned} y(t,0) &= 0 \\ \frac{\partial y}{\partial \xi}(t,0) &= 0 \end{aligned} \qquad (11.2)$$

$$\begin{aligned} \left(EI\frac{\partial^2 y}{\partial \xi^2} + c_D I\frac{\partial^3 y}{\partial \xi^2 \partial t}\right)|_{\xi=l} &= 0 \\ \frac{\partial}{\partial \xi}\left[\left(EI\frac{\partial^2 y}{\partial \xi^2} + c_D I\frac{\partial^3 y}{\partial \xi^2 \partial t}\right)\right]|_{\xi=l} &= 0 \end{aligned} \qquad (11.3)$$

and initial conditions

$$\begin{aligned} y(0,\xi) &= \Phi(\xi) \\ \frac{\partial y}{\partial t}(0,\xi) &= \Psi(\xi). \end{aligned}$$

To facilitate our discussions, we consider an <u>undamped</u> and <u>unforced</u> version (i.e., $\gamma = c_D I = 0$, $f = 0$) of the above system. Rather than force and moment balance, we consider *energy formulations* for the beam. For a segment of the beam in $[\xi, \xi + \Delta\xi]$, one can argue that the kinetic energy (at a given time t) is given by

$$KE(t) = T(t) = \frac{1}{2}\int_{\xi}^{\xi+\Delta\xi} \rho(\frac{\partial y}{\partial t}(t,\zeta))^2 d\zeta,$$

and hence the kinetic energy of the entire beam is given by

$$T(t) = \frac{1}{2}\int_0^l \rho\dot{y}^2(t,\xi)d\xi.$$

Similarly, the potential (or strain) energy U of the beam (see Section 5.2.1) at any given time t is given by

$$PE(t) = U(t) = \frac{1}{2}\int_0^l EI(\frac{\partial^2 y}{\partial \xi^2})^2 d\xi.$$

A fundamental tenet of the mechanics of rigid or elastic bodies is Hamilton's "Principle of Stationary Action" (often, in a misnomer, referred to as Hamilton's principle of "least action") which postulates that any system undergoing motion during a period $[t_0, t_1]$ will exhibit motion $y(t, \xi)$ that provides the "least action" for the system with a stationary value. The "Action" is defined by

$$A = \int_{t_0}^{t_1} (KE - PE) dt$$

$$= \int_{t_0}^{t_1} [T(t) - U(t)] dt.$$

For the beam of Example 6, this means that the motion or vibrations $y(t, \xi)$ must provide a stationary value to the action

$$A[y] = \int_{t_0}^{t_1} \int_0^l [\frac{1}{2}\rho\dot{y}^2 - \frac{1}{2}EI(y'')^2] d\xi dt$$

integrated over any time interval $[t_0, t_1]$. Through the *calculus of variations* (a field of mathematics that was the precursor to modern control theory), this leads to an equation of motion for the vibrations y that the beam motion must satisfy.

To further explore this, we consider $y(t, \xi)$ as the motion of the beam and consider a family of variations $y(t, \xi) + \epsilon\eta(t, \xi)$ where η is chosen so that $y + \epsilon\eta$ is an "admissible variation," i.e., $y + \epsilon\eta$ must satisfy the essential boundary conditions (11.2).

We define $V = H_L^2(0, l) = \{\varphi \in H^2(0, l) | \varphi(0) = \varphi'(0) = 0\}$. Let $\psi \in C^2(t_0, t_1)$ with $\psi(t_0) = \psi(t_1) = 0$. Then $\eta \in \mathcal{N} \equiv \{\eta | \eta = \psi\varphi, \varphi \in V\}$ satisfies η is C^2 in t, H^2 in ξ with $\eta(t_0, \xi) = \eta(t_1, \xi) = 0$ and $\eta(t, 0) = \eta'(t, 0) = 0$. Then by Hamilton's principle, we must have that $A[y + \epsilon\eta]$ for $\epsilon > 0, \eta \in \mathcal{N}$, possesses a stationary value at $\epsilon = 0$. That is,

$$\frac{d}{d\epsilon} A[y + \epsilon\eta]|_{\epsilon=0} = 0. \tag{11.4}$$

Since

$$A[y + \epsilon\eta] = \int_{t_0}^{t_1} \int_0^l [\frac{1}{2}\rho(\dot{y} + \epsilon\dot{\eta})^2 - \frac{1}{2}EI(y'' + \epsilon\eta'')^2] d\xi dt,$$

we find from (11.4) the weak (in time t and space ξ) form of the undamped, unforced version of equation (11.1) given by

$$0 = \int_{t_0}^{t_1} \int_0^l [\rho\ddot{y}\eta - EIy''\eta''] d\xi dt \tag{11.5}$$

for all $\eta \in \mathcal{N}$.

To explore further, we formally integrate by parts in the first term (with respect to t) to obtain

$$\int_{t_0}^{t_1} \int_0^l \rho \dot{y} \dot{\eta} \, d\xi \, dt = -\int_{t_0}^{t_1} \int_0^l \rho \ddot{y} \eta \, d\xi \, dt + \int_0^l \rho \dot{y} \eta \, d\xi \Big|_{t=t_0}^{t=t_1}$$

$$= -\int_{t_0}^{t_1} \int_0^l \rho \ddot{y} \eta \, d\xi \, dt$$

since $\eta(t_0, \xi) = \eta(t_1, \xi) = 0$. Since η has the form $\eta = \psi\varphi$, equation (11.5) has the form

$$\int_{t_0}^{t_1} \int_0^l [\rho \ddot{y} \varphi + EIy''\varphi'']\psi \, d\xi \, dt = 0 \qquad (11.6)$$

for all $\psi \in C^2[t_0, t_1]$ with $\psi(t_0) = \psi(t_1) = 0$, and all $\varphi \in V$. Since this holds for arbitrary ψ, we must have in the $L_2(t_0, t_1)$ sense

$$\int_0^l [\rho \ddot{y} \varphi + EIy''\varphi''] \, d\xi = 0 \qquad \text{for all } \varphi \in V.$$

In our former notation of Gelfand triples with $V = H_L^2(0, l)$ and $H = L_2(0, l)$, this may be written

$$\langle \rho \ddot{y}, \varphi \rangle_{V^*,V} + \langle EIy'', \varphi'' \rangle_H = 0 \qquad \text{for all } \varphi \in V$$

in the $L_2(t_0, t_1)$ sense, which is exactly the "weak" or "variational" form of the beam equation similar to those we have encountered previously in Chapter 8. Note that in fact the true variational form was given in (11.5); that is,

$$\int_{t_0}^{t_1} [-\langle \rho \dot{y}, \varphi \rangle \dot{\psi} + \langle EIy'', \varphi'' \rangle \psi] \, dt = 0$$

for all $\varphi \in V$ and $\psi \in C^2[t_0, t_1]$ with $\psi(t_0) = \psi(t_1) = 0$. (See the proofs and our remarks concerning solutions in the $L_2(t_0, t_1; V)^* \cong L_2(t_0, t_1; V^*)$ sense in the well-posedness (existence) results for second-order systems discussed earlier in Chapter 9).

We note that if the variational solution y has additional smoothness so that $y \in V \cap H^4(0, l)$ (more precisely $EIy'' \in H^2(0, l)$), then we can integrate by parts twice (with respect to ξ) in the second term of (11.6) to obtain in place of (11.6):

$$\int_{t_0}^{t_1} \int_0^l [\rho \ddot{y}\varphi + (EIy'')'']\varphi]\psi d\xi dt + \int_{t_0}^{t_1} -(EIy'')\varphi'|_{\xi=0}^{\xi=l}\psi dt$$

$$+ \int_{t_0}^{t_1} (EIy'')'\varphi|_{\xi=0}^{\xi=l}\psi dt = 0$$

for $\varphi \in V, \psi \in C^2[t_0, t_1]$ with $\psi(t_0) = \psi(t_1) = 0$. This can be written

$$\int_{t_0}^{t_1} \left[\int_0^l [\rho \ddot{y}\varphi + (EIy'')'']\varphi]d\xi + EIy''\varphi'|_{\xi=l} + (EIy'')'\varphi|_{\xi=l} \right] \psi dt = 0$$

for arbitrary $\varphi \in V$. We note once again that this results in the strong or classical form of the equations

$$\rho \frac{\partial^2 y}{\partial t^2} + \frac{\partial^2}{\partial \xi^2}(EI\frac{\partial^2 y}{\partial \xi^2}) = 0$$

with the essential boundary conditions

$$y(t, 0) = y'(t, 0) = 0$$

as well as the natural boundary conditions

$$EI\frac{\partial^2 y}{\partial \xi^2}(t, l) = \frac{\partial}{\partial \xi}(EI\frac{\partial^2 y}{\partial \xi^2})(t, l) = 0$$

holding.

We remark on an additional perspective of the weak or variational or energy formulation of dynamical systems such as the beam equation. The weak or variational form may be thought of as *Euler's equations* in the *calculus of variations*. If one wishes to minimize a cost functional

$$J(y) = \int F(t, y, \dot{y})dt = A[y],$$

then the condition of stationarity

$$\frac{d}{d\epsilon} J(y + \epsilon \eta)|_{\epsilon=0} = 0$$

implies

$$\int (F_y \eta + F_{\dot{y}}\dot{\eta})dt = 0$$

which is the "true" Euler's equation. We may then invoke the duBois-Reymond lemma [Adams, Berkovitz, Ewing, Sagan] stated below to obtain

$$\frac{\partial F}{\partial y} = \frac{d}{dt}\frac{\partial F}{\partial \dot{y}}$$

in a *distributional* sense. Specifically, we have

Lemma 11.1 *(duBois-Reymond Lemma)*
If

$$\int (G_1\eta + G_2\dot{\eta}) = 0$$

for all η, then

$$\frac{d}{dt}G_2 = G_1$$

in a weak sense. In other words, G_1 is the distributional derivative of G_2.

If we assume enough smoothness and integrate by parts, we obtain the strong form of Euler's equation:

$$-\frac{d}{dt}F_{\dot{y}}(t, y, \dot{y}) + F_y(t, y, \dot{y}) = 0.$$

In the derivation above, we considered the undamped and unforced version of the equation. In the case of the forced beam, we can add a conservative force term $W = fy$ in our derivation, and we will obtain the desired $\langle f, \varphi \rangle$ term in our result. However, there is no known way to derive the weak form with damping. In other words, Hamilton's principle is essentially valid for *conservative forces*, but it does not conveniently handle nonconservative (dissipative) forces (damping).

12 Finite Element Approximations and the Trotter-Kato Theorems

12.1 Finite Elements

We will now consider finite element approximations or Galerkin approximations for general first-order systems for which parabolic systems are special cases. Consider

$$\begin{cases} \dot{x}(t) = Ax + F & \text{in } V^* \ (H \text{ if possible}) \\ x(0) = x_0 \end{cases} \tag{12.1}$$

where $V \hookrightarrow H \hookrightarrow V^*$ is the usual Gelfand triple. We can write the above system in the weak or variational form (i.e., in V^*) as

$$\begin{cases} \langle \dot{x}(t), \varphi \rangle_{V^*,V} + \sigma(x(t), \varphi) = \langle F(t), \varphi \rangle_{V^*,V} \\ x(0) = x_0 \end{cases}$$

for $\varphi \in V$. If σ is V continuous and V-elliptic, and $S(t) \sim e^{At}$ (i.e., A is the infinitesimal generator of a C_0 semigroup $S(t)$), we can write

$$x(t) = S(t)x_0 + \int_0^t S(t-\xi)F(\xi)d\xi \tag{12.2}$$

where $x \in L_2(0,T;V) \bigcap C(0,T;H)$ and $\dot{x} \in L_2(0,T;V^*)$. We can use this formulation to give a nice treatment of finite element approximations of Galerkin type.

In general, this is an infinite dimensional space; therefore, we want to project the system into a finite dimensional space in which we can compute. Let $H^N = span\{B_1^N, B_2^N, ..., B_N^N\} \subset V$ be the approximation of H. Typical choices for the B_j^N are piecewise linear or piecewise cubic splines (e.g., see [BK]). The idea is to replace (12.1) by

$$\begin{cases} \dot{x}^N(t) = A^N x^N(t) + F^N(t) & \text{in } H^N \\ x^N(0) = x_0^N, \end{cases}$$

or equivalently, replace (12.2) by

$$x^N(t) = S^N(t)x_0^N + \int_0^t S^N(t-\xi)F^N(\xi)d\xi$$

where $S^N(t) = e^{A^N t}$, i.e., A^N is the generator of $S^N(t)$.

One of the key constructs we need is $P^N : H \to H^N$ which is called the *orthogonal projection* of H onto H^N. In other words, P^N is defined by

$$\langle P^N \varphi - \varphi, \psi \rangle = 0 \qquad \text{for all } \psi \in H^N$$

or

$$|P^N \varphi - \varphi|_H = \inf_{\psi \in H^N} |\psi - \varphi|_H.$$

We would like $F^N \to F$ and $x_0^N \to x_0$, so we take $x_0^N = P^N x_0$ and $F^N(t) = P^N F(t)$. We also want $A^N \in \mathcal{L}(H^N)$ and $A^N \approx A$. However, we have defined $S^N(t) = e^{A^N t}$ and $S(t) \sim e^{At}$; therefore, if we had $S^N(t) \to S(t)$, then we would be able to argue $x^N(t) \to x(t)$. This is considered in the Trotter-Kato theorem which will be discussed in the next chapter.

Now to relate this to the computational aspects of finite elements, we first restrict the equations

$$\langle \dot{x}(t), \varphi \rangle + \sigma_1(x(t), \varphi) = \langle F(t), \varphi \rangle \qquad \text{for all } \varphi \in V \qquad (12.3)$$

to $H^N \times H^N$. In other words, let

$$x^N(t) = \sum_{j=1}^{N} w_j^N(t) B_j^N$$

be a trial solution with

$$x^N(0) = \sum_{j=1}^{N} w_{0j}^N B_j^N.$$

Substituting this into (12.3), we have

$$\langle \sum_{j=1}^{N} \dot{w}_j^N(t) B_j^N, \varphi \rangle + \sigma_1(\sum_{j=1}^{N} w_j^N(t) B_j^N, \varphi) = \langle F(t), \varphi \rangle \qquad (12.4)$$

for $\varphi \in H^N$. Successively choose $\varphi = B_1^N, B_2^N, ...B_N^N$ in (12.4). From this we obtain an $N \times N$ vector system for $w^N(t) = (w_1^N(y), .., w_N^N(t))^T$ given by

$$\sum_{j=1}^{N} \dot{w}_j^N(t) \langle B_j^N, B_i^N \rangle + \sum_{j=1}^{N} w_j^N(t) \sigma(B_j^N, B_i^N) = \langle F(t), B_i^N \rangle \qquad (12.5)$$

for $i = 1, 2, ..., N$.

Using standard engineering and applied mathematics terminology, we define the *mass* matrix $M^N = (\langle B_i^N, B_j^N \rangle)$, the *stiffness* matrix $K^N = (\sigma(B_i^N, B_j^N))$, and the column vector $F^N(t) = (\langle F(t), B_i^N \rangle)$. Then (12.5) can be written

$$\begin{cases} M^N \dot{w}^N(t) + K^N w^N(t) = F^N(t) \\ w^N(0) = w_0^N \end{cases} \qquad (12.6)$$

or

$$\begin{cases} \dot{w}^N(t) = -(M^N)^{-1} K^N w^N(t) + (M^N)^{-1} F^N(t) \\ w^N(0) = w_0^N. \end{cases}$$

Considering w_0^N, we have $x^N(0) = P^N x_0$ which implies $\langle P^N x_0 - x_0, B_i^N \rangle = 0$ for $i = 1, ..., N$. However, $x_0^N = \sum_{j=1}^N w_{0j}^N B_j^N$. Therefore,

$$\langle \sum_{j=1}^N w_{0j}^N B_j^N - x_0, B_i^N \rangle = 0$$

for $i = 1, ..., N$ which gives

$$\sum_{j=1}^N w_{0j}^N \langle B_j^N, B_i^N \rangle = \langle x_0, B_i^N \rangle.$$

Defining $w_0^N = col(w_{01}^N, ..., w_{0N}^N)$, we have

$$w_0^N = (M^N)^{-1} col(\langle x_0, B_i^N \rangle).$$

From this, our system for w becomes

$$\begin{cases} \dot{w}^N(t) = -(M^N)^{-1} K^N w^N(t) + (M^N)^{-1} F^N(t) \\ w_0^N = (M^N)^{-1} col(\langle x_0, B_i^N \rangle). \end{cases}$$

However, we normally do not solve the system in this form. If $\langle B_i, B_j \rangle = 0$ for $i \neq j$, then M^N is diagonal and the system of the form (12.6) is an easier system with which to work. More generally, the (finite element) system is solved in the form

$$\begin{aligned} M^N \dot{w}^N(t) &= -K^N w^N(t) + F^N(t) \\ M^N w_0^N &= col(\langle x_0, B_i^N \rangle), \end{aligned}$$

rather than inverting the matrix M^N which is, if not diagonal, usually a banded matrix (tri-banded for piecewise linear splines, seven banded for cubic splines, etc.—see [BK, StrangFix]), lending itself to fast algebraic solvers (e.g., those based on LU decompositions).

Convergence of finite element approximations for partial differential equations is often carried out in the context of weak formulations while corresponding arguments for delay systems use an operator approach. Both of these rely on versions of the Trotter-Kato theorems which we give in the next section.

12.2   Trotter-Kato Approximation Theorem

The Trotter-Kato Approximation Theorem is the functional analysis operator version of the Lax Equivalence Principle used in finite element and finite difference approximations for PDE's which dates back to the 1960s. The fundamental ideas underlying the Lax Equivalence Principle are that "consistency" and "stability" are achieved if and only if we have "convergence" of our system. If we have a first-order PDE (we could also motivate with a second-order in time PDE)

$$y_t = Ay$$

and approximations

$$y_t^N = A^N y^N,$$

then consistency refers to $A^N \to A$ in some sense. Stability refers to $|e^{A^N t}| \le Me^{\omega t}$, and convergence means $e^{A^N t} \to e^{At}$ in some sense. For relevant and much more detailed discussions, see [RM].

There are two different versions of the Trotter-Kato theorem which we will discuss here. We will first consider the *operator convergence* form of the Trotter-Kato theorem.

Theorem 12.1 *Let X and X^N be Hilbert spaces such that $X^N \subset X$. Let $P^N : X \to X^N$ be an orthogonal projection of X onto X^N. Assume $P^N x \to x$ as $N \to \infty$ for all $x \in X$. Let A^N, A be infinitesimal generators of C_0 semigroups $S^N(t), S(t)$ on X^N, X respectively satisfying*

(i) there exists M, ω such that $|S^N(t)| \le Me^{\omega t}$ for each N

(ii) there exists $\mathcal{D}$ dense in X such that for some λ, $(\lambda I - A)\mathcal{D}$ is dense in X and $A^N P^N x \to Ax$ for all $x \in \mathcal{D}$.

Then for each $x \in X$, $S^N(t)P^N x \to S(t)x$ uniformly in t on compact intervals $[0, T]$.

The arguments for this result are given in [Pa, Chapter 3, Theorem 4.5]. We next will examine the *resolvent convergence* version of the Trotter-Kato theorem. This form is a modification of the previous version.

Theorem 12.2 *Replace* (*ii*) *in the above theorem by* (*ĩi*) *defined as*

(*ĩi*) *There exists* $\lambda \in \rho(A) \cap\limits_{N=1}^{\infty} \rho(A^N)$ *with* $\text{Re}(\lambda) > \omega$ *so that* $R_\lambda(A^N)P^N x \to R_\lambda(A)x$ *for each* $x \in X$.

Then under this and the remaining conditions in the above theorem, the conclusions also hold.

For proofs, see [Pa, Chap. 3, Theorems 4.2, 4.3, 4.4]. See also [BK, Theorem 1.14]. Convergence rates are given in [BK, Theorem 1.16].

We can also formulate a version (Theorem 12.4 below) of the Trotter-Kato theorem that is readily applied in the setting of sesquilinear forms and Gelfand triples $V \hookrightarrow H \hookrightarrow V^*$ with approximating subspaces $H^N \subset V$. This will be useful in certain problems since it is not necessary for $H^N \subset \mathcal{D}(A)$ where $\mathcal{D}(A)$ carries both the essential and natural boundary conditions. We may need to only choose an appropriate approximation H^N such that $H^N \subset V$ which carries just the essential boundary conditions. If we restrict ourselves to first-order systems in the context of the Gelfand triple $V \hookrightarrow H \hookrightarrow V^*$ with H^N approximating V as $N \to \infty$, then we have a special case of the Trotter-Kato approximation theorem for such first-order systems.

Let the condition (**C1**) be given by

(**C1**) For each $z \in V$, there exists $\hat{z}^N \in H^N$ such that $|z - \hat{z}^N|_V \to 0$ as $N \to \infty$.

Suppose σ is V-elliptic, i.e., $\text{Re } \sigma(\varphi, \varphi) \geq \delta|\varphi|_V^2$ for $\delta > 0$. Also assume σ is V continuous, i.e., $|\sigma(\varphi, \psi)| \leq \gamma|\varphi|_V|\psi|_V$. Let $P^N : H \to H^N$ be an orthogonal projection. Then

$$|P^N z - z| = \inf\{|z^N - z|_H \mid z^N \in H^N\}.$$

Under (**C1**), we have $|P^N z - z|_H \leq |\hat{z}^N - z|_H \leq |\hat{z}^N - z|_V \to 0$ as $N \to \infty$. Therefore, under (**C1**), $P^N z \to z$ for $z \in H$.

Next we define A^N. We have $\sigma(\varphi, \psi) = \langle -A\varphi, \psi\rangle_H$ for $\varphi \in (\mathcal{D})(A) = \{\psi \in V|A\psi \in H\}, \psi \in V$ where A is an infinitesimal generator of a C_0 semigroup of contractions on H, i.e., $|e^{At}| \leq 1$. We define A^N through the restriction of σ to $H^N \times H^N$. Therefore, $A^N : H^N \to H^N$ is defined by

$$
\begin{aligned}
\sigma(\varphi^N, \psi^N) &= \langle -A^N \varphi^N, \psi^N \rangle_H \quad \varphi^N, \psi^N \in H^N \\
&= \langle -\mathcal{A} \varphi^N, \psi^N \rangle_{V^*, V}.
\end{aligned}
$$

By V-ellipticity, we obtain A^N is an infinitesimal generator of a contraction semigroup (see Theorem 6.1) on H^N, i.e., $|e^{A^N t}| \leq 1$.

Theorem 12.3 *If σ is V-elliptic, V continuous, and* **(C1)** *holds, then*

$$
R_\lambda(A^N) P^N z \to R_\lambda(A) z
$$

in the V norm for $z \in H$ and $\lambda = 0$.

Proof

Let $z \in H$ and take $\lambda = 0$. Now define $w^N = R_\lambda(A^N) P^N z$ and $w = R_\lambda(A) z$ where $\sigma(\varphi, \psi) = \langle -A\varphi, \psi \rangle_H, \varphi \in \mathcal{D}(A), \psi \in V$. By definition, we have $w \in \mathcal{D}(A)$.

By (C1), there exists $\hat{w}^N \in H^N$ such that $|\hat{w}^N - w|_V \to 0$ as $N \to \infty$. Let $z^N = w^N - \hat{w}^N$. We need to show $z^N \to 0$ in V.

Since $R_\lambda(A) = (\lambda I - A)^{-1}$, we have

$$
\begin{aligned}
\sigma(w, z^N) &= \langle -A R_\lambda(A) z, z^N \rangle_H \\
&= \langle z, z^N \rangle
\end{aligned}
$$

while

$$
\begin{aligned}
\sigma(w^N, z^N) &= \langle -A^N R_\lambda(A^N) z, z^N \rangle_H \\
&= \langle z, z^N \rangle,
\end{aligned}
$$

and hence these are equal. Thus

$$
\begin{aligned}
\delta |z^N|_V^2 &\leq \sigma(z^N, z^N) \\
&= \sigma(w^N, z^N) - \sigma(\hat{w}^N, z^N) \\
&= \sigma(w, z^N) - \sigma(\hat{w}^N, z^N) \\
&= \sigma(w - \hat{w}^N, z^N) \\
&\leq \gamma |w - \hat{w}^N|_V |z^N|_V.
\end{aligned}
$$

Therefore, we have $\delta |z^N|_V \leq \gamma |w - \hat{w}^N|_V$. However, $|\hat{w}^N - w|_V \to 0$ implies $|z^N|_V \to 0$ and thus $|w - w^N|_V \leq |w - \hat{w}^N|_V + |\hat{w}^N - w^N|_V \to 0$.

Remark: Theorem 2.2 of [BI] is a parameter dependent version of this, i.e., $A = A(q), A^N = A^N(q^N), q, q^N \in Q$, which can be used in inverse problems. See [BI] for additional discussions.

Theorem 12.4 *Suppose σ is V-elliptic, V continuous, and* **(C1)** *holds. Then $S^N(t)P^N z \rightarrow S(t)z$ in the V norm for each $z \in H$ uniformly in t on compact intervals.*

Proof The arguments involve using Tanabe-type estimates (Chapter 6) after considering an appropriate setting and using the Trotter-Kato theorem, first in H and then in V. Let $X = H, X^N = H^N, P^N : H \rightarrow H^N$ be an orthogonal projection. Then $P^N z \rightarrow z$ for all $z \in H$ by **(C1)**. Convergence in the H norm follows immediately from application of the Trotter-Kato theorem in H. To obtain V convergence is somewhat more work and more delicate; for these details we refer the reader to [BI, Theorem 2.3].

13 Delay Systems: Linear and Nonlinear

The study of delay equations has a long history in fields as disparate as economics [FrischHolme] and ecology [Hutch], with some early applications in engineering found in research concerning the stability of naval vessels [Minorsky]. Furthermore, there has also been extensive use of delay equations in modeling biological systems and indeed both May's and Murray's classic texts ([May] and [Murray], respectively) have significant sections devoted to delay equations. For the interested reader, there are solid introductory texts [BellmanCooke, Driver] (including those that focus heavily on applications to biological systems [Cushing, Kuang]) and thorough (if somewhat theoretical) advanced texts [Diekmannetal, Goreckietal, JKH3].

Since delay systems, like partial differential equations, are infinite dimensional state systems, the development of underlying theory and computational ideas has depended heavily on functional analytic formulations. In particular, semigroups have played a fundamental role in both conceptual and computational contributions. To illustrate some of these, we turn to two examples of delayed systems which can be written in abstract form so that the semigroup and general finite approximation ideas of the previous sections are directly applicable. For the first example, we return to Example 3 on delayed lethal effects of pesticides to motivate formulation of approximation methods for general linear delay systems. We then present in some detail cellular level HIV models illustrating nonlinear delay systems as well as systems with cumulative delays, i.e., a continuum of distributed delays represented by a hysteretic kernel. These represent what are commonly called functional differential equations (FDE).

13.1 Linear Delay Systems and Approximation

We return to Example 3 where the dynamics were given by the delay system

$$\frac{da}{dt}(t) = \int_{-7}^{-5} n(t+\tau)k(\tau)d\tau - (d_a + p_a)\,a(t)$$

$$\frac{dn}{dt}(t) = ba(t) - (d_n + p_n)\,n(t) - \int_{-7}^{-5} n(t+\tau)k(\tau)d\tau \qquad (13.1)$$

$$a(\theta) = \Phi(\theta), \quad n(\theta) = \Psi(\theta), \quad \theta \in [-7,0)$$

$$a(0) = a^0, \qquad n(0) = n^0,$$

where k is a probability density kernel which we assumed has the property $k(\tau) \geq 0$ for $\tau \in [-7, -5]$ and $k(\tau) = 0$ for $\tau \in (-\infty, -7) \cup (-5, 0]$.

Here $a(t)$ and $n(t)$ denote the number of adults and neonates, respectively, in the population at time t. We lumped the mortality due to insecticide into one parameter p_a for the adults, p_n for the neonates, and denoted by d_a and d_n the background or natural mortalities for adults and neonates, respectively. The parameter b is the rate at which neonates are born into the population.

We found in Section 3.6.3 that this system was a special case (for $n = 2$) of the general system

$$\dot{z}(t) = L(z_t), \quad (z(0), z_0) = (\eta, \phi), \qquad (13.2)$$

where $z_t(\xi) = z(t + \xi)$, $\xi \in [-r, 0]$ and

$$L(z_t) = \sum_{i=0}^{m} A_i z(t - \tau_i) + \int_{-r}^{0} K(\xi)z(t + \xi)d\xi,$$

where $0 = \tau_0 < \tau_1 < \cdots < \tau_m = r$, and A_i and $K(\xi)$ are $n \times n$ matrices. For general functions $\phi \in C(-r, 0; \mathbb{R}^n)$, the mapping $L(\cdot) : C(-r, 0; \mathbb{R}^n) \to \mathbb{R}^n$ has the form

$$L(\phi) = \sum_{i=0}^{m} A_i \phi(-\tau_i) + \int_{-r}^{0} K(\xi)\phi(\xi)d\xi. \qquad (13.3)$$

Letting $x(t) = (z(t), z_t) \in X$, $X \equiv \mathbb{R}^n \times L_2(-r, 0; \mathbb{R}^n)$, where the Hilbert space X has the inner product

$$\langle (\eta, \phi), (\zeta, \psi) \rangle_X = \langle \eta, \zeta \rangle_{\mathbb{R}^n} + \int_{-r}^{0} \langle \phi(\xi), \psi(\xi) \rangle_{\mathbb{R}^n} d\xi, \qquad (13.4)$$

we found that equation (13.2) can be formulated as

$$\dot{x}(t) = \mathcal{A}x(t)$$

$$x(0) = x_0, \qquad (13.5)$$

where $x_0 = (\eta, \phi)$ is the initial condition. Here the linear operator $\mathcal{A} : \mathcal{D}(\mathcal{A}) \subset X \to X$ with domain

$$\mathcal{D}(\mathcal{A}) = \{(\varphi(0), \varphi) \in X \,|\, \varphi \in H^1(-r, 0; \mathbb{R}^n)\} \tag{13.6}$$

is given by

$$\mathcal{A}(\varphi(0), \varphi) = (L(\varphi), \varphi'). \tag{13.7}$$

The system of functional differential or delay equations described in (13.2) can be simulated using an algorithm first developed by Banks and Kappel for linear systems [BKap1]. To use this algorithm, one must first convert the system to an abstract evolution equation (AEE) (as described above and in Section 3.6.3) with corresponding solution semigroup $S(t) : X \to X$ and then approximate the solutions in a space spanned by piecewise linear splines. Thus one can numerically calculate the generalized Fourier coefficients of the approximate solution in the spline basis representation and recover an approximation to the solution of (13.2).

Define X^N to be an approximating subspace [BKap1, BK] of X. In particular, we choose $X^N = X_1^N$ to be the piecewise linear spline subspaces of X discussed in detail in [BKap1]. We briefly outline the results for the piecewise linear subspaces X_1^N (see Section 4 of [BKap1]) given by

$$X_1^N = \{(\phi(0), \phi) \,|\, \quad \phi \text{ is a continuous first-order spline function}$$
$$\text{with knots at } t_j^N = -jr/N, j = 0, 1, \dots, N\}.$$

Let P^N be the orthogonal projection of X onto X^N, and $\mathcal{A}^N$ as the approximating operator for $\mathcal{A}$ given by $\mathcal{A}^N = P^N \mathcal{A} P^N$. Then the problem given in (13.5) is approximated by the finite dimensional problem

$$\begin{aligned} \dot{x}^N(t) &= \mathcal{A}^N x^N(t), \quad t \geq 0, \\ x^N(0) &= P^N x_0. \end{aligned} \tag{13.8}$$

We fix the basis X_1^N for a special case of X^N corresponding to the partition $t_j^N = -j(r/N)$ for $j = 0, \dots, N$. Then the basis is defined by

$$\hat{\beta}^N = (\beta^N(0), \beta^N) \text{ where } \beta^N = (e_0^N, e_1^N, \dots, e_N^N) \otimes \mathbb{I}_n, \tag{13.9}$$

and the e_j^N's are piecewise linear splines defined by

$$e_j^N(t_i^N) = \delta_{ij} \text{ for } i, j = 0, 1, \dots, N.$$

When $\mathcal{A}^N$ is restricted to X_1^N, we have a matrix representation A_1^N of $\mathcal{A}^N$. Define $w^N(t)$ so that $x^N(t) = \hat{\beta}^N w^N(t)$. Then solving for $x^N(t)$ in the finite dimensional system (13.8) is equivalent to solving for $w^N(t)$ in the linear vector system

$$
\begin{aligned}
\dot{w}^N(t) &= A_1^N w^N(t) \\
w^N(0) &= w_0^N,
\end{aligned}
\tag{13.10}
$$

where $\hat{\beta}^N w_0^N = P^N x_0$. We note that having obtained $w^N(t)$, the product $\hat{\beta}^N w^N(t)$ converges uniformly in t to the solution $x(t) = (z(t), z_t)$ of (13.5) if we can argue the convergence $x^N(t) \to x(t)$. To do this we use the Trotter-Kato theorem.

We follow the mathematical theory first given in [BKap1] for proofs which use the operator convergence form of the Trotter-Kato theorem, Theorem 12.1. This theory guarantees that the solutions of (13.8) converge to the solution of (13.5) as the number N of evenly supported elements in $[-r, 0]$ goes to infinity. The convergence arguments can be carried equivalently in either X or X^N since they are topologically equivalent.

First, for the linear splines with corresponding projections P^N : $X_g \to X^N = X_1^N$, we have that $P^N x \to x$ for all $x \in X$. This follows from standard spline estimates [BKap1, BK, Schultz]. Next we use the fact that X and X_g (see Section 3.6.3) are topologically equivalent and the arguments of Section 3.6.3 to argue

$$
\begin{aligned}
\langle \mathcal{A}^N x, x \rangle_{X_g} &= \langle P^N A P^N x, x \rangle_{X_g} \\
&\leq \langle A P^N x, P^N x \rangle_{X_g} \leq \omega \mid P^N x \mid^2_{X_g} \\
&\leq \tilde{\omega} \mid x \mid^2_{X_g},
\end{aligned}
\tag{13.11}
$$

where $\tilde{\omega}$ is independent of N. It then follows from the range arguments in Section 3.6.3 plus the Lumer-Phillips theorem that $\mathcal{A}^N \in G(1, \tilde{\omega})$ in X_g and hence $\mathcal{A}^N \in G(M, \tilde{\omega})$ in X where M is independent of N. Thus condition (i) of Theorem 12.1 is satisfied. To obtain the desired convergence we only need argue that condition {ii} of Theorem 12.1 is satisfied. But the arguments in [BKap1] reveal that for

$$
\mathcal{D} \equiv \{ \hat{\phi} = (\phi(0), \phi) \in X \mid \phi \in C[-r, 0], \dot{\phi}(0) = L(\phi) \}
$$

one finds that both $\mathcal{D}$ and, for some $\lambda > 0$, $(\mathcal{A} - \lambda I)\mathcal{D}$ are dense in X and X_g. Moreover, the arguments in [BKap1] (based on $P^N x \to x$ and continuity of L) also yield $\mathcal{A}^N P^N x \to \mathcal{A} x$ for all $x \in \mathcal{D}$. Thus by the operator convergence form of the Trotter-Kato Theorem 12.1, we have $x^N(t) \to x(t)$ uniformly on compact intervals $[0, T]$.

13.2 Modeling of Viral Delays in HIV Infection Dynamics

We next consider classes of functional or delay differential equation models which arise in attempts to describe temporal delays in HIV pathogenesis. In particular, we consider models for incorporating arbitrary variability (i.e., general probability distributions) for these delays into systems that cannot readily be reduced to a finite number of coupled ordinary differential equations (as is done in the method of stages discussed below). Based on the derivations in [BBH], we introduce several classes of nonlinear models (including discrete and distributed delays), and present some functional analysis based discussions of theoretical and computational approaches. The models we discuss have been used and validated with experimental data in successful inverse problem efforts. This has been confirmed by statistical significance tests for the presence of delays and the importance of including delays [BBH]. We also found that the models are quite sensitive to the mean of the distribution which describes the delay in viral production, whereas the variance of this distribution has relatively little impact.

We discuss in some detail the underlying biology since this is germane to the functional form of the equations and hence the functional analytic approach required to treat the resulting systems. Viruses are obligate intra-cellular parasites with a multitude of pathways for infecting and reproducing within their target hosts. The human immunodeficiency virus (HIV) is a lentivirus that is the etiological agent for the slow, progressive, and fatal acquired immunodeficiency syndrome (AIDS) for which there is currently no known cure.

For HIV, the core of the virus is composed of single-stranded viral RNA and protein components. As depicted in Figure 13.1, when an HIV virion comes into contact with an uninfected target cell, the viral envelope glycoproteins fuse to the cell's lipid bilayer at a CD4 receptor site and the viral core is injected into the cell. Once inside, the protein components enable transcription and integration of the viral RNA into viral DNA and then incorporation into the cellular DNA (provirus). With its altered cellular DNA, the cell produces capsids and protein envelopes and transcribes multiple copies of viral RNA. The cell assembles a virion by then encasing the viral RNA in a capsid followed by a protein envelope. The new HIV virion pushes out through the cell membrane budding off in chains of virions (though sometimes single virions do float away into the plasma). Clearly the time from

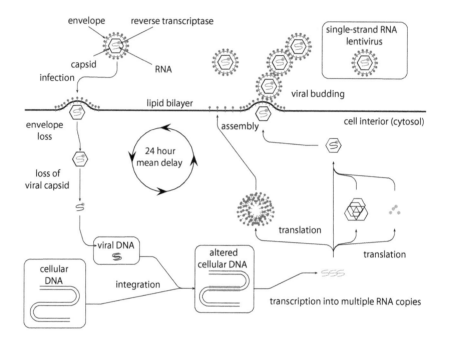

Figure 13.1: HIV infection pathway.

viral infection to viral production (sometimes called the *eclipse phase* [NelsonMittlerPerelson]) is not instantaneous, and (as indicated in the figure) it is estimated that the first viral release occurs approximately 24 hours after the initial infection [MitSulNeumPerel].

Within the HIV modeling community, there is considerable debate upon the proper compartment definitions. The multi-compartment model introduced and employed in [BBH] describes pathways from the moment a virion contacts the appropriate receptor site as the beginning of *acute infection*. If the acutely infected cell survives through its first viral release, roughly 3 hours later the physiological characteristics of the cell change and it is subsequently classified as a *chronically infected* cell. Note that in the chronic stage, it is possible for the cells to continue to divide and to produce virions, albeit at a much slower rate than acutely infected or non-infected cells.

One classical approach to numerically simulating these systems with delays is sometimes referred to as the *method of stages* and is described in [CoxMiller, Jensen, Lloyd, Lloyd2, MitSulNeumPerel, NelsonPerelson]. All of these papers which include delays in their models represent them using a gamma distribution (or Erlang distribution [Jensen]) to describe a distribution of delays, and thus can reduce the resulting system of integro-differential equations to a system of (non-delayed) ordinary differential equations [Murray] (we shall return to this in more detail later). This non-delayed system can easily be simulated using standard numerical integration techniques (e.g., Runge-Kutta) in standard mathematical software. An alternative method (a discussion of which is presented in Section 13.1 for linear systems) that first converts a delay system into an abstract evolution equation (before numerical simulation) is discussed in Section 3.6.3 and [Banks82, BBu2, BKap1]. This approach allows for simulation of systems with general delay kernels describing the delay distributions, and does not require that the model be *reduced* to a system of non-delayed ODEs (it is only approximated by ODEs).

In the course of developing our model, we employ a delay to mathematically represent the temporal lag between the initial viral infection and the first release of new virions. We concentrate on the mathematical modeling of viral dynamics, focusing in particular on the mathematical aspects and biological nature of the delays in primary infection. The models are extensions of previous modeling work on HIV infection dynamics for *in vitro* laboratory experiments from the (continuous) delay differential equations developed in [Bortz, et al.], which in turn were based on a discrete dynamical system from [HolteEmerman]. We used this model elsewhere [BBH] along with the *in vitro* data (see Figure 13.2) from [RogelWuEmerman] to illustrate the methods we discuss for both forward and inverse problems. Our primary interest here is to illustrate the functional equations and functional analytical methodology required when treating cellular level data containing significant variability.

We begin with a modification of the system of ordinary differential equations developed in [Bortz, et al.] given by

$$
\begin{aligned}
\dot{V}(t) &= -cV(t) + n_A A(t) + n_C C(t) - pV(t)T(t) \\
\dot{A}(t) &= (r_v - \delta_A - \gamma - \delta X(t))A(t) + pV(t)T(t) \\
\dot{C}(t) &= (r_v - \delta_C - \delta X(t))C(t) + \gamma A(t) \\
\dot{T}(t) &= (r_u - \delta_u - \delta X(t) - pV(t))T(t) + S ,
\end{aligned}
\tag{13.12}
$$

for $0 \le t \le t_f$ with t_f finite, where the parameters and the compartments are described in Tables 13.1 and 13.2, respectively, and t is the continuous independent time variable. In the first equation, the

Table 13.1: *in vitro* model parameters

Notation	Description
c	Infectious viral clearance rate
n_A	Infectious viral production rate for acutely infected cells
n_C	Infectious viral production rate for chronically infected cells
γ	Rate at which acutely infected cells become chronically infected
r_v	Birth-rate for virally infected cells
r_u	Birth-rate for uninfected cells
δ_A	Death-rate for acutely infected cells
δ_C	Death-rate for chronically infected cells
δ_u	Death-rate for uninfected cells
δ	Density dependent overall cell death-rate
p	Rate of infection
S	Constant rate of target cell replacement

$-pV(t)T(t)$ term is designed to account for the biological fact that upon infecting a cell, a virion is unable to infect additional target cells. Models possessing this term are inherently different from many *in vivo* models in which the (large) number of target cells is assumed to be constant. If over the time scale of interest, the T variable were a constant T_0 (such as in [MitSulNeumPerel, NelsonMurrayPerelson]), the equation would be

$$\dot{V}(t) = -(c + pT_0)V(t) + n_A A(t) + n_C C(t)$$

and we could define a new coefficient $c' = c + pT_0$ for the $V(t)$ term. For our *in vitro* model, we do not have this situation, as the target cell population is not replenished and thus not held constant in the experiment. In computational results (reported in [BBH]), we omitted the $-pV(t)T(t)$ term from the first equation of (13.12) and were also able to attain reasonable fits for our limited data set (albeit with different parameters in the models) along with statistically significant results analogous to those reported in [BBH].

We call attention in (13.12) to the terms describing the rates of change in the population of virions, acutely infected cells, chronically infected cells, and uninfected cells ($\dot{V}(t)$, $\dot{A}(t)$, $\dot{C}(t)$, and $\dot{T}(t)$, respec-

Table 13.2: *in vitro* Model Compartments

Notation	Description
V	Infectious viral population
A	Acutely infected cells
C	Chronically infected cells
T	Uninfected or target cells
X	Total cell population (infected and uninfected) $(A + C + T)$

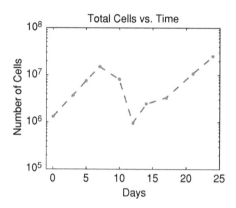

Figure 13.2: Log plot of experimental data (10 observations) from [RogelWuEmerman].

tively, in (13.12)). In particular, we should comment on the form of the nonlinear terms (e.g., $pV(t)T(t)$). Terms such as $pV(t)T(t)$ are obviously only first approximations to the density dependent (on V and T) component of the rate of new infections. A more realistic model requires that this term, dependent on both $V(t)$ and $T(t)$, be bounded in the limit, i.e., saturation should be modeled in the nonlinear term so that in the limit it is (at least) affine in V or T. For well-posedness considerations, the term pVT is more appropriately replaced by a function $p(V,T)$ where $x \mapsto p(x)$, $x = (V,T)$ is globally Lipschitz (see [Banks82] for the standard form of this assumption). This will be described in more detail later; however, for our initial purposes in modeling discussions, the simpler term will suffice.

As mentioned earlier, it is known that there exist temporal delays between viral infection and viral production and between productive acute infection and chronic infection.

A central focus of the modeling effort in [BBH] was on attempting to obtain reasonable mathematical representations of these delays. The problem of how to mathematically represent these phenomena is decidedly nontrivial and includes issues such as how to account for intra-individual variability (e.g., intercellular variability arising within a single infected individual or laboratory assay) and/or inter-individual variability arising between individual subjects or data from multiple assays. Here we do not specifically address or distinguish these different sources of variability, although the model is sufficiently flexible to account for either type. These issues are highly significant and dealing with the levels of variability and the resulting mathematical ramifications is of primary interest.

Let the delay in the first equation in (13.12) be modeled by treating the delay time τ between acute infection and viral production as a probabilistic quantity (i.e., a random variable) with distribution $P_1(\tau)$ so that the first equation in (13.12) is replaced by (see below for a more detailed discussion of the foundations underlying such an equation)

$$\dot{V}(t) = -cV(t)+n_A \int_{-\infty}^{0} A(t+\tau)dP_1(\tau)+n_C C(t)-pV(t)T(t). \quad (13.13)$$

Likewise, let the delay between acute infectivity and chronic infectivity (with distribution $P_2(\tau)$) be represented in altered forms of the second and third equations of (13.12) by

$$
\begin{aligned}
\dot{A}(t) &= (r_v - \delta_A - \delta X(t))A(t) - \gamma \int_{-\infty}^{0} A(t+\tau)dP_2(\tau) \\
&\quad + pV(t)T(t) \qquad\qquad\qquad\qquad\qquad\qquad (13.14) \\
\dot{C}(t) &= (r_v - \delta_C - \delta X(t))C(t) + \gamma \int_{-\infty}^{0} A(t+\tau)dP_2(\tau). \quad (13.15)
\end{aligned}
$$

Note that assuming Dirac distributions δ_a with atoms a at $a_1 = -\tau_1 < 0$ and $a_2 = -\tau_1 - \tau_2 < -\tau_1 < 0$, respectively, for P_1, P_2 reduces the system to

$$
\begin{aligned}
\dot{V}(t) &= -cV(t) + n_A A(t - \tau_1) + n_C C(t) - pV(t)T(t) \\
\dot{A}(t) &= (r_v - \delta_A - \delta X(t))A(t) - \gamma A(t - \tau_1 - \tau_2) + pV(t)T(t) \\
\dot{C}(t) &= (r_v - \delta_C - \delta X(t))C(t) + \gamma A(t - \tau_1 - \tau_2) \\
\dot{T}(t) &= (r_u - \delta_u - \delta X(t) - pV(t))T(t) + S\,,
\end{aligned}
$$

$$(13.16)$$

for $0 \le t \le t_f$ where t_f is finite. Moreover it becomes the special case

$$
\begin{aligned}
\dot{V}(t) &= -cV(t) + n_A \int_{-\infty}^{0} A(t+\tau)k_1(\tau)d\tau + n_C C(t) - pV(t)T(t) \\
\dot{A}(t) &= (r_v - \delta_A - \delta X(t))A(t) - \gamma \int_{-\infty}^{0} A(t+\tau)k_2(\tau)d\tau + pV(t)T(t) \\
\dot{C}(t) &= (r_v - \delta_C - \delta X(t))C(t) + \gamma \int_{-\infty}^{0} A(t+\tau)k_2(\tau)d\tau \\
\dot{T}(t) &= (r_u - \delta_u - \delta X(t) - pV(t))T(t) + S,
\end{aligned}
$$

(13.17)

whenever P_1, P_2 possess probability densities k_1, k_2, respectively. In the discussions in [BBH], all numerical simulations for each of the systems of functional differential equations (FDE) given above were performed using the linear spline based methods described below. There it was found, not surprisingly, the presence of nonzero delays has a dramatic effect upon the simulations. Issues relating to the exact nature of τ and whether or not it should be modeled as a fixed value for each cell or distributed across cells and how this distribution can be represented, are the focus of discussions below. Further evidence of the statistical significance of the presence of the delay is given in [BBH].

13.3 Nonlinear Delay Systems

In order to approximate solutions to the systems described in (13.16) and (13.17), one may first convert them to an AEE and then approximate in a space spanned by piecewise linear (or even higher order) splines (i.e., in a Galerkin approach, which is equivalent to a linear finite element approximation in partial differential equations). One is then able to numerically calculate the generalized Fourier coefficients of approximate solutions relative to the splines, and with these coefficients, recover an approximation to the solutions of (13.16) or (13.17).

We turn to the mathematical aspects of these nonlinear FDE systems and present an outline of the necessary mathematical and numerical analysis foundations. First we describe the conversion of the nonlinear FDE system to an abstract evolution equation (AEE) system as well as provide existence and uniqueness results for a solution to the FDE. We then provide a fundamental approximation framework including convergence results.

For nonlinear delay systems such as those discussed above, approximation in the context of a linear semigroup framework as presented so far is not direct, but one can use the ideas of that theory as a basis for a wide class of nonlinear delay system approximations. Details in this direction can be found in the early work [Banks79, Banks82, Kappel82] which is a direct extension of the results in [BBu1, BBu2, BKap1] to nonlinear delay systems.

We summarize an extension of well-developed theory given in [Banks79, Banks82] most pertinent to our efforts here. We do this in the context of the motivating HIV models of the previous section. Note that when we consider the discrete delay system (13.16) using Dirac distributions as P_1 and P_2, we take the delays $\tau_1, \tau_2 > 0$ and without loss of generality assume $r > \tau_1 + \tau_2 > 0$ is finite throughout.

Let

$$z(t) = (z_1(t), z_2(t), z_3(t), z_4(t))^T = (V(t), A(t), C(t), T(t))^T$$

and

$$z_t(\theta) = z(t + \theta), \ -r \leq \theta \leq 0, \ r \in \mathbb{R}^+.$$

Our system, as described in (13.16) or (13.17), can then be written as

$$\dot{z}(t) = L(z(t), z_t) + f_1(z(t)) + f_2(t) \qquad \text{for } 0 \leq t \leq t_f$$

$$(z(0), z_0) = (\Phi(0), \Phi) \in X, \ \Phi \in \mathscr{C}(-r, 0; \mathbb{R}^4)$$

(13.18)

where t_f is finite and for $(\eta, \phi) \in \mathbb{R}^4 \times \mathscr{C}(-r, 0; \mathbb{R}^4) \subset X \equiv \mathbb{R}^4 \times L_2(-r, 0; \mathbb{R}^4)$,

$$
L(\eta, \phi) = \begin{bmatrix} -c & 0 & n_C & 0 \\ 0 & r_v - \delta_A & 0 & 0 \\ 0 & 0 & r_v - \delta_C & 0 \\ 0 & 0 & 0 & r_u - \delta_u \end{bmatrix} \eta
$$

$$
+ \ n_A \left[\delta_{(1,2)}\right]_{(4,4)} \int_{-r}^0 \phi(\theta) \, dP_1(\theta)
$$

$$
+ \gamma \left(\left[\delta_{(3,2)}\right]_{(4,4)} - \left[\delta_{(2,2)}\right]_{(4,4)} \right) \int_{-r}^0 \phi(\theta) \, dP_2(\theta) \ ,
$$

$$
f_1(\eta) = \begin{bmatrix} -p\eta_1\eta_4 \\ -\delta \left(\sum_{i=2}^4 \eta_i \right) \eta_2 + p\eta_1\eta_4 \\ -\delta \left(\sum_{i=2}^4 \eta_i \right) \eta_3 \\ -\delta \left(\sum_{i=2}^4 \eta_i \right) \eta_4 - p\eta_1\eta_4 \end{bmatrix} ,
$$

$$f_2(t) = [0, 0, 0, S]^T, \ 0 \leq t \leq t_f.$$

Here $\left[\delta_{(i,j)}\right]_{(4,4)}$ denotes a 4 by 4 matrix with a one in the (i, j)th element and zeros elsewhere. Moreover, Φ is the initial time history of the system on $[-r, 0]$, $\{P_1, P_2\}$ are probability distributions, and X is the state space, a Hilbert space with the usual inner product

$$\langle (\eta, \phi), (\zeta, \psi) \rangle_X = \eta^T \zeta + \int_{-r}^{0} \phi(\theta)^T \psi(\theta) \, d\theta,$$

for (η, ϕ), $(\zeta, \psi) \in X$. The presentation in [Banks79, BBH] can be followed to obtain local existence and uniqueness of a solution to (13.18).

As noted earlier, the nonlinearities exemplified by terms such as $p z_1 z_4$ are biologically unrealistic. However, these nonlinear terms in f_1 can be replaced by standard saturation limited nonlinearities such as

$$p z_1 z_4 \quad \text{by} \quad p_1(z_1) z_4$$
$$\text{and} \quad \delta z_i z_j \quad \text{by} \quad \delta_i(z_i) z_j \quad (\text{for } i, j = 2, 3, 4),$$

where

$$p_1(z_1) = \begin{cases} 0 & z_1 < 0 \\ p z_1 & 0 \le z_1 \le \bar{z}_1 \\ p \bar{z}_1 & \bar{z}_1 < z_1, \end{cases} \tag{13.19}$$

and

$$\delta_i(z_i) = \begin{cases} 0 & z_i < 0 \\ \delta z_i & 0 \le z_i \le \bar{x}_i \\ \delta \bar{z}_i & \bar{z}_i < z_i, \end{cases} \tag{13.20}$$

(for finite upper bounds $\bar{z}_i \in \mathbb{R}^+, i = 1, 2, 3, 4$), and where f_1 can be replaced by

$$\tilde{f}_1(\eta) = \begin{bmatrix} -p_1(\eta_1)\eta_4 \\ -\left(\sum_{i=2}^{4} \eta_i\right)\delta_2(\eta_2) + p_1(\eta_1)\eta_4 \\ -\left(\sum_{i=2}^{4} \eta_i\right)\delta_3(\eta_3) \\ -\left(\sum_{i=2}^{4} \eta_i\right)\delta_4(\eta_4) - p_1(\eta_1)\eta_4 \end{bmatrix}, \quad \eta \in \mathbb{R}^4. \tag{13.21}$$

Note that p_i and δ_i are globally bounded functions satisfying $p_1(z_1) \le p\bar{z}_1$ and $\delta_i(z_i) \le \delta\bar{z}_i$. Indeed they are differentiable, satisfying $p_1'(z_1) \le p$ and $\delta_i'(z_i) \le \delta$.

One can establish the global existence and uniqueness of a solution to

$$\dot{z}(t) = L(z(t), z_t) + \tilde{f}_1(x(t)) + f_2(t) \qquad \text{for } 0 \le t \le t_f,$$
$$(z(0), z_0) = (\Phi(0), \Phi) \in X \tag{13.22}$$

with t_f finite, in a number of ways including Picard iterates or fixed point arguments (uniform contraction principle, Schauder Fixed Point—see [JKH4]). To pursue this, we first define $F = L + \tilde{f}_1$ on X by

$$F(\eta, \phi) = L(\eta, \phi) + \tilde{f}_1(\eta) .$$

Given that all pertinent components of the equation (13.22) are differentiable, we can conclude that the function F is differentiable and thus have

Lemma 13.1 *The function $F = L + \tilde{f}_1 : X \to \mathbb{R}^4$ is differentiable.*

Lemma 13.2 *For all $(\eta, \phi), (\zeta, \psi) \in X$ the function $F = L + \tilde{f}_1$ satisfies a global Lipschitz condition*

$$|F(\eta, \phi) - F(\zeta, \psi)| \leq K_L \{|\eta - \zeta| + |\phi - \psi|\} \tag{13.23}$$

for some fixed constant $K_L > 0$.

A rather complete treatment of the mathematical aspects of the associated approximate inverse problems (including a convergence analysis) can be found in [BanksBortzIP]. Stability analysis related to the approximations for delay systems can be found in [BBu2, BKap1], while for general stability results for delay systems, see [BellmanCooke].

Remark 13.1 *Well-posedness and approximation results for the systems above can actually be obtained from the arguments for nonautonomous nonlinear delay systems in [Banks79]. In that approach, one requires all discrete delays to appear in the linear part of the system dynamics while continuously distributed delays may appear in the nonlinear part. One then writes the system dynamics as an autonomous linear part plus a nonlinear perturbation. The linear part generates a linear semigroup as in [BBu1, BBu2, BKap1] and as discussed earlier in this text. One then uses the linear semigroup in a variation of parameters implicit representation of the solution to the nonlinear system. Mathematical tools used then are Picard iterates for existence, and the Trotter-Kato convergence theorem (for the linear semigroups) plus a Gronwall inequality. An alternative (and more general) approach given in [Banks82] and generalized in the next section eschews use of the Trotter-Kato theorem in treating general nonautonomous nonlinear delay systems which allows discrete delays in the nonlinear components* *of the system. As we shall see below, the main mathematical tools are dissipative properties of the general nonlinear operator $\mathcal{A}$ representing the system and direct approximation $\mathcal{A}^N(t) \to \mathcal{A}(t)$ (no Trotter-Kato!) along with a Gronwall inequality. Finally, in another alternative, a nonlinear semigroup (with dissipative generators) approach along with a corresponding nonlinear Trotter-Kato convergence results are given by Kappel in [Kappel82]. With these results one can treat directly general autonomous nonlinear delay systems in the spirit of the linear semigroup approach given in this text.*

13.4 State Approximation and Convergence for Nonlinear Delay Systems

We consider the general system

$$\dot{z}(t) = f(t, z(t), z_t, z(t - \tau_1), \ldots, z(t - \tau_m)) + f_2(t), \quad 0 \le t \le T,$$
(13.24)

$$z_0 = \phi,$$

where $f = f(t, \eta, \psi, y_1, \ldots, y_m) : [0, T] \times X \times \mathbb{R}^{nm} \to \mathbb{R}^n$. Here $X = \mathbb{R}^n \times L_2(-r, 0; \mathbb{R}^n)$, $0 < \tau_1 < \ldots < \tau_m = r$, z_t denotes the usual function $z_t(\theta) = z(t + \theta)$, $-r \le \theta \le 0$, and $\phi \in H^1(-r, 0)$. We shall make use of the following hypotheses throughout our presentation.

(H1f) The function f satisfies a global Lipschitz condition:

$$
\begin{aligned}
|f(t, \eta, \psi, y_1, \ldots, y_m) &- f(t, \xi, \tilde{\psi}, w_1, \ldots, w_m)| \\
&\le K \left(|\eta - \xi| + |\psi - \tilde{\psi}| + \sum_{i=1}^{m} |y_i - w_i| \right)
\end{aligned}
$$

for some fixed constant K and all $(\eta, \psi, y_1, \ldots, y_m)$, $(\xi, \tilde{\psi}, w_1, \ldots, w_m)$ in $X \times \mathbb{R}^{nm}$ uniformly in t.

(H2f) The function $f : [0, T] \times X \times \mathbb{R}^{nm} \to \mathbb{R}^n$ is differentiable.

Remark 13.2 *If we define the function $F : [0, T] \times \mathbb{R}^n \times C(-r, 0; \mathbb{R}^n) \subset X \to \mathbb{R}^n$ given by*

$$F(t, x) = F(t, \eta, \psi) = f(t, \eta, \psi, \psi(-\tau_1), \ldots, \psi(-\tau_m))$$
(13.25)

we observe that even though f satisfies (H1f), F will not satisfy a continuity hypothesis on its domain in the X norm.

We define the nonlinear operator $\mathcal{A}(t) : \mathcal{D}(\mathcal{A}) \subset X \to X$ by

$$\mathcal{D}(\mathcal{A}) \equiv \{(\psi(0), \psi) \mid \psi \in H^1(-r, 0)\}$$

$$\mathcal{A}(t)(\psi(0), \psi) = (F(t, \psi(0), \psi), \psi').$$

Theorem 13.1 *Assume that (H1f) holds and let $x(t; \phi, f_2) = (z(t; \phi, f_2), z_t(\phi, f_2))$, where z is the solution of (13.24) corresponding to $\phi \in H^1$, $f_2 \in L_2$. Then for $\zeta = (\phi(0), \phi)$, $x(t; \phi, f_2)$ is the unique solution on $[0, T]$ of*

$$x(t) = \zeta + \int_0^t [\mathcal{A}(\sigma)x(\sigma) + (f_2(\sigma), 0)]d\sigma. \qquad (13.26)$$

Furthermore, $f_2 \to x(t; \phi, f_2)$ is weakly sequentially continuous from L_2 (with weak topology) to X (with strong topology).

Again these results can be established in one of several ways including fixed point theorem arguments or Picard iteration arguments. Either of these approaches can be used to establish existence, uniqueness and continuous dependence of the solution of (13.26). For existence, uniqueness and continuous dependence of the solution of (13.24), we note that our condition (H1f) is a global version of the hypothesis of Kappel and Schappacher in [KapSch], so that in the autonomous case their results also yield immediately the desired result for (13.24).

The uniqueness of solutions to (13.26) follows in the usual manner once we establish that $\mathcal{A}$ satisfies a dissipative inequality. Indeed, defining a weighting function g exactly as in [BKap1, p. 500] and in Section 3.6.3 and the corresponding weighted inner product $\langle \, , \, \rangle_{X_g} = \langle \, , \, \rangle_g$ on X, one can show without difficulty that (H1f) implies the dissipative inequality (see [Barbu, p. 71]) for the nonlinear operator $\mathcal{A}(t)$

$$\langle \mathcal{A}(t)x - \mathcal{A}(t)w, x - w \rangle_g \le \omega \langle x - w, x - w \rangle_g \qquad (13.27)$$

for all $x, w \in \mathcal{D}(\mathcal{A})$ and all t.

Turning next to the approximation of (13.24) through approximation of (13.26), we let $X^N = X_1^N$ be the linear spline subspaces of X discussed in detail in [BKap1] and in Section 13.1. A careful study of the arguments behind our presentation reveals that the approximation results given here hold for general spline approximations. For example, if one were to treat cubic spline approximations (X_3^N of [BKap1]), one would use the appropriate approximation analogues of Theorem

2.5 of [Schultz] and Theorem 21 of [SchuVarg] (e.g., see Theorem 4.5 of [Schultz]). Hereafter, when we write X^N, the reader should understand that we mean X_1^N of [BKap1] and Section 13.1.

Let $P^N = P_g^N$ be the orthogonal projection (in $\langle\,,\,\rangle_g$) of X onto X^N so that as we have already discussed it immediately follows that $P^N x \to x$ for all $x \in X$. Similar to the approach in [BKap1] as extended in [Banks82], we define the approximating operator $\mathcal{A}^N(t) = P^N A(t) P^N$ and consider the approximating equations in X^N given by

$$x^N(t) = P^N \zeta + \int_0^t [\mathcal{A}^N(\sigma) x^N(\sigma) + P^N(f_2(\sigma), 0)] d\sigma \qquad (13.28)$$

which, because X^N is finite-dimensional, are equivalent to

$$\dot{x}^N(t) = \mathcal{A}^N(t) x^N(t) + P^N(f_2(t), 0), \quad x^N(0) = P^N \zeta. \qquad (13.29)$$

From (13.27) and the definition of $\mathcal{A}^N$ in terms of the self-adjoint projections P^N, we have at once that under (H1f) the sequence $\{\mathcal{A}^N\}$ satisfies on X a uniform dissipative inequality

$$\langle \mathcal{A}^N(t)x - \mathcal{A}^N(t)w, x - w \rangle_g \leq \omega \langle x - w, x - w \rangle_g \qquad (13.30)$$

Uniqueness of solutions of (13.28) then follows immediately from this inequality. Upon recognition that (13.29) is equivalent to a nonlinear ordinary differential equation in Euclidean space with the right-hand side satisfying a global Lipschitz condition, one can easily argue existence of solutions for (13.29) and hence for (13.28) on any finite interval $[0, T]$. Our main result to be discussed here, which insures that solutions of (13.29) converge to those of (13.26), can now be stated.

Theorem 13.2 *Assume (H1f), (H2f). Let $\zeta = (\phi(0), \phi)$, $\phi \in H^1$ and $f_2 \in L_2(0, T)$ be given, with x^N and x the corresponding solutions on $[0, T]$ of (13.29) and (13.26), respectively. Then $x^N(t) \to x(t) = (z(t; \phi, f_2), z_t(\phi, f_2))$, as $N \to \infty$, uniformly in t on $[0, T]$.*

Remark 13.3 *One can actually obtain slightly stronger results than those given in Theorem 13.2. One can consider solutions of (13.26) and (13.29) corresponding to initial data $(z(0), z_0) = (\eta, \phi) = \zeta$ with $\eta \in \mathbb{R}^n$, $\phi \in L_2$ (i.e., $\zeta \in X$) and argue that the results of Theorem 13.2 hold also in this case.*

To indicate briefly our arguments for Theorem 13.2, we consider for given initial data ζ and perturbation f_2 the corresponding solutions x and x^N of (13.26) and (13.28). Defining $\Delta^N(t) \equiv x^N(t) - x(t)$ and $F_2(t) = (f_2(t), 0)$, we obtain immediately that

$$\Delta^N(t) = (P^N - I)\zeta + \int_0^t \left[\mathcal{A}^N(\sigma)x^N(\sigma) - \mathcal{A}(\sigma)x(\sigma) + (P^N - I)F_2(\sigma) \right] d\sigma.$$

$$(13.31)$$

We next use a rather standard technique for analysis of differential equations (see [Barbu]), the foundations of which we state as a lemma since we shall refer to it again.

Lemma 13.3 *If X is a Hilbert space and $x : [a, b] \to X$ is given by*

$$x(t) = x(a) + \int_a^t y(\sigma)d\sigma,$$

then

$$|x(t)|^2 = |x(a)|^2 + 2 \int_0^t \langle x(\sigma), y(\sigma) \rangle d\sigma.$$

This lemma is essentially a restatement of the well-known result [Barbu, p. 100] that in a Hilbert space

$$\frac{d}{dt}\left(\frac{1}{2}|x(t)|^2 \right) = \langle \dot{x}(t), x(t) \rangle.$$

Applying Lemma 13.3 to (13.31), we obtain

$$|\Delta^N(t)|^2 = |(P^N - I)\zeta|^2 + 2\int_0^t \langle \mathcal{A}^N(\sigma)x^N(\sigma)$$

$$-\mathcal{A}(\sigma)x(\sigma) + (P^N - I)F_2(\sigma), \Delta^N(\sigma) \rangle d\sigma$$

$$= |(P^N - I)\zeta|^2 + 2\int_0^t \langle \mathcal{A}^N(\sigma)x^N(\sigma) - \mathcal{A}^N(\sigma)x(\sigma), \Delta^N(\sigma) \rangle d\sigma$$

$$+2\int_0^t \langle (\mathcal{A}^N(\sigma) - \mathcal{A}(\sigma))x(\sigma) + (P^N - I)F_2(\sigma), \Delta^N(\sigma) \rangle d\sigma.$$

If we use (13.30) on the first integral term in this last expression, we then have

$$
\begin{aligned}
|\Delta^N(t)|^2 \;\leq\;& |(P^N - I)\zeta|^2 + 2\int_0^t \omega |\Delta^N(\sigma)|^2 d\sigma \\
&+ 2\int_0^t \langle (\mathcal{A}^N(\sigma) - A(\sigma))x(\sigma) + (P^N - I)F_2(\sigma), \Delta^N(\sigma)\rangle d\sigma \\
\leq\;& |(P^N - I)\zeta|^2 + 2\int_0^t \omega |\Delta^N(\sigma)|^2 d\sigma \\
&+ 2\int_0^t \Big\{ \tfrac{1}{2}|(\mathcal{A}^N(\sigma) - A(\sigma))x(\sigma)|^2 + \tfrac{1}{2}|\Delta^N(\sigma)|^2 \\
&+ \tfrac{1}{2}\big|(P^N - I)F_2(\sigma)\big|^2 + \tfrac{1}{2}\big|\Delta^N(\sigma)\big|^2 \Big\} d\sigma \\
=\;& |(P^N - I)\zeta|^2 + \int_0^t \big|(\mathcal{A}^N(\sigma) - A(\sigma))x(\sigma)\big|^2 d\sigma \\
&+ \int_0^t \big|(P^N - I)F_2(\sigma)\big|^2 d\sigma + 2(\omega + 1)\int_0^t |\Delta^N(\sigma)|^2 d\sigma.
\end{aligned}
$$

An application of Gronwall's inequality to this then yields the estimate

$$
|\Delta^N(t)|^2 \leq [\epsilon_1(N) + \epsilon_2(N) + \epsilon_3(N)] \exp\left(2(\omega + 1)t\right), \qquad (13.32)
$$

where

$$
\epsilon_1(N) = |(P^N - I)\zeta|^2,
$$

$$
\epsilon_2(N) = \int_0^T \big|(\mathcal{A}^N(\sigma) - A(\sigma))x(\sigma)\big|^2 d\sigma,
$$

$$
\epsilon_3(N) = \int_0^T \big|(P^N - I)F_2(\sigma)\big|^2 d\sigma.
$$

Since $P^N \to I$ strongly in X and the convergence $|(P^N - I)F_2(\sigma)| \to 0$ in ϵ_3 is dominated, to prove Theorem 13.2 it suffices to argue that $\epsilon_2(N) \to 0$ as $N \to \infty$. To that end, we state the following sequence of lemmas.

Lemma 13.4 *Assume (H1f) and let $X \equiv \{x = (\phi(0), \phi) \mid \phi \in H^2\}$. Then for each t, $\mathcal{A}^N(t)x \to A(t)x$ as $N \to \infty$ for each $x \in X$.*

Lemma 13.5 *Let $\tilde{C} \equiv \{(\zeta, f_2) \in \mathcal{D}(A) \times L_2(0, T) \mid \phi \in H^2, f_2 \in H^1$, with $\dot{\phi}(0) = F(\zeta) + f_2(0)$ where $\zeta = (\phi(0), \phi)\}$. Assume that (H1f), (H2f) hold. Then for $(\zeta, f_2) \in \tilde{C}$ the corresponding solution $\sigma \to x(\sigma) = (z(\sigma), z_\sigma)$ of (13.26) (z is the solution of (13.24)) satisfies $x(\sigma) \in X$ for each $\sigma \in [0, T]$.*

Lemma 13.6 *Assume (H1f), (H2f) and let $(\zeta, f_2) \in \tilde{\mathcal{C}}$ with x^N and x the corresponding solutions of (13.28) and (13.26). Then $x^N(t) \to x(t)$ uniformly in t on $[0, T]$.*

Lemma 13.7 *Assume (H1f). Then the solutions of (13.26) and (13.28) depend continuously (in the $X \times L_2$ topology) on $(\zeta, f_2) \in \mathcal{D}(\mathcal{A}) \times L_2$, uniformly in t on $[0, T]$.*

Lemma 13.8 *The set $\tilde{\mathcal{C}}$ defined in Lemma 13.5 is dense in $X \times L_2$.*

We obtain the convergence of Theorem 13.2 by combining Lemmas 13.6, 13.7, and 13.8. The proof of Lemma 13.7 employs Lemma 13.3 along with Gronwall's inequality in much the same way as above in deriving (13.32) from (13.31). We note that Lemma 13.5 requires hypothesis (H2f) (this is the only place in which it is used) in order to obtain enough smoothness of solutions x of (13.26) so that $x(\sigma) \in \mathcal{X}$ for each σ.

In developing the estimates to establish Lemma 13.6 (which, by our above remarks, requires only that we argue $\epsilon_2(N) \to 0$), we use heavily the standard spline estimates found in [Schultz] and [SchuVarg]. Lemmas 13.4 and 13.5 yield that $\mathcal{A}^N(\sigma)x(\sigma) \to \mathcal{A}(\sigma)x(\sigma)$ for each σ so that to prove Lemma 13.6 one only need show that this convergence is dominated, thereby guaranteeing $\epsilon_2(N) \to 0$. In making the arguments for Lemma 13.6, one obtains at the same time error estimates on the convergence in Theorem 13.2. For example, one readily finds the following: for $\phi \in H^2$, f satisfying (H1f), (H2f), $\dot{\phi}(0) = F(\phi(0), \phi)$ and $f_2 \equiv 0$, the convergence $x^N(t) \to x(t)$ is $O(1/N)$. For higher-order splines and higher-order convergence estimates (e.g., cubic splines with convergence $O(1/N^3)$), one of course needs additional smoothness (beyond (H2f)) on f.

The convergence given in Theorem 13.2 yields state approximation techniques for nonlinear FDE systems based on the spline methods developed in [BKap1]. These results can be applied directly to control and identification problems, the latter of which are discussed in [Banks82].

13.5   Fixed Delays versus Distributed Delays

We return to the cellular level HIV models discussed earlier in this chapter. If we assume that the delays τ_1, τ_2 are fixed for each cell, then we can precisely describe the capacity of each member of the population (of infected cells) to produce virions as a function of time. In other words, exactly τ_1 units of time after a cell becomes infected, it begins producing virus. Exactly τ_2 units of time later, that same cell then becomes chronically infected (assuming it lives to this stage).

Such a system can be obtained from (13.13)-(13.15) along with the equation for $\dot{T}$ by assuming Heaviside measures (yielding Dirac distributions) P_1, P_2 with unit jumps at $-\tau_1 < 0$ and $-\tau_1 - \tau_2 < 0$. This corresponds to Dirac delta "densities" and results in the system

$$
\begin{aligned}
\dot{V}(t) &= -cV(t) + n_A \int_{-\infty}^{0} A(t+\tau)\,\delta_{-\tau_1}(\tau)\,d\tau + n_C C(t) - pV(t)\,T(t) \\
\dot{A}(t) &= (r_v - \delta_A - \delta X(t))\,A(t) - \gamma \int_{-\infty}^{0} A(t+\tau)\,\delta_{-\tau_1-\tau_2}(\tau)\,d\tau \\
&\quad + pV(t)\,T(t) \\
\dot{C}(t) &= (r_v - \delta_C - \delta X(t))\,C(t) + \gamma \int_{-\infty}^{0} A(t+\tau)\,\delta_{-\tau_1-\tau_2}(\tau)\,d\tau \\
\dot{T}(t) &= (r_u - \delta_u - \delta X(t) - pV(t))\,T(t) + S ,
\end{aligned}
$$

$$(13.33)$$

which is exactly the system (13.16).

In order to overcome the (biologically untenable) assumption that each cell begins producing virus at a fixed time after infection, a number of authors have used a gamma function (just one example of a number of distributions which could be used to model this process) as the distribution for the time to viral production of infected cells. In particular, the use of the scaled gamma function

$$
k_\Gamma(\xi; b, n) = \frac{\xi^{n-1}}{(n-1)!\,b^n} \tag{13.34}
$$

as the distribution of the times to viral production (for infected cells) is a particularly popular modeling choice. The advantage to using this gamma distribution is that with a clever change of variables, the distributed delay system can be rewritten as a non-delayed system of ODEs (and thus easily simulated using standard software packages, e.g., MATLAB®). This transformation is known as the *method of stages*, and a full derivation of the equivalent system of ODEs (where just the viral production delay is modeled with a gamma distribution) is presented in [MitSulNeumPerel] (for a treatment of the method of stages for delay systems, also see [Murray]). These dynamics are easily illustrated in terms of the $\dot{V}$ equation. By introducing the internal variables $y = (y_1, \ldots, y_n)^T$, the first equation of (13.17) reduces to the equivalent system

$$
\begin{aligned}
\dot{V}(t) &= -cV(t) + n_C C(t) + n_A y_n(t) - pV(t)T(t) \\
\dot{y}_1(t) &= (A(t) - y_1(t))/b \\
\dot{y}_2(t) &= (y_1(t) - y_2(t))/b \\
&\vdots \quad \vdots \quad \vdots \\
\dot{y}_n(t) &= (y_{n-1}(t) - y_n(t))/b \,.
\end{aligned}
\tag{13.35}
$$

If we write the vector version of $\dot{y}$, we find that

$$
\dot{y}(t) = (B_1 + B_2)\, y(t) + \tfrac{1}{b}
\begin{pmatrix}
A(t) \\
0 \\
\vdots \\
0
\end{pmatrix}
$$

$$
= (B_1 + B_2)\, y(t) + F(t) \,,
$$

where

$$
B_1 = \operatorname{diag}\left(-b^{-1}\right), \quad
B_2 =
\begin{pmatrix}
0 & \cdots & \cdots & \cdots & 0 \\
1/b & \ddots & & & \vdots \\
0 & \ddots & \ddots & & \vdots \\
\vdots & \ddots & \ddots & \ddots & \vdots \\
0 & \cdots & 0 & 1/b & 0
\end{pmatrix} \,.
$$

Note that if we were to use this method, an analogous system of equations is needed to represent the delays in (13.14) and (13.15). Furthermore, while this kernel does generate equations which are simple to simulate on a computer (an unnecessary simplification, given the well-developed numerical methods for delay systems developed in [BBu1, BKap1] and described in Section 13.1, Remark 13.1, and Section 13.4), the resulting model is *equivalent* to a model with *completely identical deterministic internal dynamics* with constant parameters b and n for each subpopulation of cells. That is, the choice of the gamma kernel to describe the delay distribution yields a system that is equivalent to a completely deterministic system.

A primary advantage of the models such as (13.17) and the numerical schemes illustrated here lies in the fact that for systems that are linear in the delay term, we can consider arbitrary kernels and not just ones based upon the gamma function. Moreover, for distributions parameterized by their mean μ and variance σ^2, this flexibility will allow one to (in theory) independently identify μ and σ^2 (a feature not readily available with the gamma — see remarks below).

In order to use the method of stages (MOS) to simulate a system like (13.17), it is required that the distribution for any delay be represented by a gamma function, i.e., the use of the MOS is impossible without this assumption. Moreover, for the gamma function, the mean μ and the variance σ^2 are parameterized by b and n ($\mu = nb$ and $\sigma^2 = nb^2$), where b is a real and n is a positive integer. In the MOS implementation, n then corresponds to the order of the system of approximating ODEs (see equation (13.35)), while b is a coefficient in the system. In other words, in order to use the MOS to identify μ (using the given data in a nonlinear least squares (NLS) parameter identification framework), it is necessary to identify the *number* of equations to be used in the approximation. This is a most challenging problem, since the NLS optimization would thus be trying to identify two real numbers using a real number and an integer. Moreover, in any iterative procedure, the order of the underlying system could change with each iteration. In summary, while the gamma distribution may well be an attractive choice in forward simulations since it leads to a simple ordinary differential equations formulation, its use in inverse problems where the estimation of the means and variances are of primary interest, is clearly not advantageous as other computational issues arise (see [Bortz02]). A treatment for more general distributions is given in [BBH].

13.5.1 First Principles Modeling of Distributed Delays

Finally in this section we present a brief derivation from first principles (with assumptions based on the biology) that supports the mathematical form in treating the delays as stochastic or random variables. Readers not interested may skip to the next section. However, the ideas are relevant in many other population (particles, cells, individuals) and disease progression models.

Let us first consider the delay between initial acute infection and initial chronic infection of a cell. It is biologically unrealistic to expect an entire population of cells to simultaneously change infection characteristics $\bar{\mu}_2$ ($\bar{\mu}_2 > 0$) hours after initial viral infection. Therefore, suppose that the delay between initial acute infection and chronic infection varies across the cell population (thus mathematically characterizing the intercellular variability) according to a probabilistic distribution

$\bar{P}_2$ with density $\bar{k}_2$. We denote by $C(t;\tau)$ the subpopulation consisting of chronically infected cells that either maintained their acute infection characteristics for τ time units or are the progeny of those same cells. In other words, for some $\tau > 0$, there exists a subpopulation $C(t;\tau)$ of the chronically infected cells which either spent τ hours as acutely infected cells (before converting to chronically infected cells) or are descendants of cells that spent exactly τ hours as acutely infected cells. Thus, the rate of change in this subpopulation of cells is governed by

$$\dot{C}(t;\tau) = (r_v - \delta_C - \delta X(t))C(t;\tau) + \gamma A(t-\tau),$$

where

$$X(t) = A(t) + C(t) + T(t)$$

and the expected value of the population of chronic cells is given by integrating with respect to the distribution $\bar{P}_2$, over admissible delays, obtaining

$$C(t) = \mathcal{E}_2[C(t;\tau)] = \int_0^\infty C(t;\tau)\bar{k}_2(\tau)d\tau. \tag{13.36}$$

Therefore, the rate of change in the total population of chronic cells is governed by

$$\begin{aligned} \dot{C}(t) &= \mathcal{E}_2[\dot{C}(t;\tau)] \\ &= (r_v - \delta_C - \delta X(t))C(t) + \gamma \int_0^\infty A(t-\tau)\bar{k}_2(\tau)d\tau \\ C(0) &= C_0, \end{aligned}$$
$$\tag{13.37}$$

where C_0 is the initial condition for the total chronically infected cell population.

Next, we consider the delay between viral infection and viral production for the acutely infected cells $A(t)$. Again, it is unreasonable to expect the entire population of acutely infected cells to simultaneously commence viral production $\bar{\mu}_1$ ($\bar{\mu}_1 > 0$) hours after infection. We suppose that the delay between infection and production (for acutely infected cells $A(t)$) varies across the population with probability distribution $\bar{P}_1$ and corresponding density $\bar{k}_1$. We also partition the expected total viral population V into those virions V_A produced by acutely infected cells and those virions V_C produced by chronically infected cells so that

$$V = V_A + V_C.$$

We then denote by $V_A(t; \tau)$ the subpopulation of virus which are produced by an acutely infected cell τ hours after being infected. Thus, the rate of change in this subgroup of virions is governed by

$$\dot{V}_A(t; \tau) \;=\; -cV_A(t; \tau) + n_A A(t - \tau) - pV_A(t; \tau)T(t) \,.$$

To obtain the (expected) number of virus at time t that have been produced by acutely infected cells, we must integrate with respect to the distribution $\bar{P}_1$, over all admissible delays

$$V_A(t) = \mathcal{E}_1[V_A(t; \tau)] = \int_0^\infty V_A(t; \tau)\bar{k}_1(\tau)d\tau \,,$$

which yields the governing equation for this larger subpopulation of virions

$$
\begin{aligned}
\dot{V}_A(t) \;&=\; \mathcal{E}_1[\dot{V}_A(t; \tau)] \\
&=\; -cV_A(t) + n_A \int_0^\infty A(t - \tau)\bar{k}_1(\tau)d\tau - pV_A(t)T(t) \,.
\end{aligned}
$$

To account for the chronically infected cells as a source of virions, we denote by V_C the subpopulation of virions produced by chronically infected cells. Thus the equation describing the rate of change in the size of this subpopulation is

$$\dot{V}_C(t) \;=\; -cV_C(t) + n_C C(t) - pV_C(t)T(t) \,,$$

where the expected value C of the total population of chronically infected cells is defined in (13.36). Therefore, the governing equations for the total population of virus are

$$
\begin{aligned}
\dot{V}(t) \;&=\; \mathcal{E}_1[\dot{V}_A(t; \tau) + \dot{V}_C(t)] \\
&=\; -c(V_A(t) + V_C(t)) + n_A \int_0^\infty A(t - \tau)\bar{k}_1(\tau)d\tau + n_C C(t) \\
&\quad -\; p(V_A(t) + V_C(t))T(t) \\
&=\; -cV(t) + n_A \int_0^\infty A(t - \tau)\bar{k}_1(\tau)d\tau + n_C C(t) - pV(t)T(t) \\
V(0) \;&=\; V_0 \,,
\end{aligned}
$$

where V_0 is the initial condition for the total virion population.

Moreover, we assume that the A and T subclasses have no subpopulation structures, and are therefore governed by

$$
\begin{aligned}
\dot{A}(t) &= (r_v - \delta_A - \delta X(t)) A(t) - \gamma \int_0^\infty A(t-\tau) \bar{k}_2(\tau) \, d\tau \\
&\quad + pV(t) T(t) \quad\quad\quad\quad\quad\quad\quad\quad\quad\quad (13.38) \\
A(0) &= A_0 \\
\dot{T}(t) &= (r_u - \delta_u - \delta X(t) - pV(t)) T(t) + S \\
T(0) &= T_0,
\end{aligned}
$$

with initial conditions A_0 and T_0. Note that in (13.38), the rate term with the delay (representing the delayed conversion of A to C) is simply the negative of the delay rate term in (13.37).

Finally, we make the change of variables $k_i(\xi) = \bar{k}_i(-\xi)$ so that the densities are now defined on $(-\infty, 0)$ instead of $(0, \infty)$ (we do this to be consistent with the notation of Section 1.5 which is standard in the FDE literature), and obtain the system

$$
\begin{aligned}
\dot{V}(t) &= -cV(t) + n_A \int_{-\infty}^0 A(t+\tau) k_1(\tau) \, d\tau + n_C C(t) - pV(t) T(t) \\
\dot{A}(t) &= (r_v - \delta_A - \delta X(t)) A(t) - \gamma \int_{-\infty}^0 A(t+\tau) k_2(\tau) \, d\tau + pV(t) T(\\
\dot{C}(t) &= (r_v - \delta_C - \delta X(t)) C(t) + \gamma \int_{-\infty}^0 A(t+\tau) k_2(\tau) \, d\tau \\
\dot{T}(t) &= (r_u - \delta_u - \delta X(t) - pV(t)) T(t) + S,
\end{aligned}
$$

which is identical to (13.17).

14 Weak* Convergence and the Prohorov Metric in Inverse Problems

We turn next to discuss functional analysis concepts (in particular the concept of weak* convergence introduced in Section 4.2) that are integral to several types of inverse problems. These problems entail parameter estimation using output observations where the parameters to be estimated are probability distributions. Such problems can be readily found in certain areas of applications [BBi, BBH, BBKW, BBPP, BD, BDTR, GRD-FP, BDEHAD, GRD-FP2, BF, BFPZ, BG1, BG2, BGIT]. Here we consider two types of measure dependent problems: individual dynamics and aggregate dynamics.

Many inverse problems involve individual dynamics and individual data. For example, in Example 5 of Section 1.7 if one had individual longitudinal data d_{ij} corresponding to the structured population density $v(t_i, \xi_j; g)$, where v is the solution to (1.22)-(1.23), corresponding to an (individual) cohort of fish all with the same individual growth rate $\frac{d\xi}{dt} = g(t, \xi)$, one could then formulate a standard ordinary least squares (OLS) problem for estimation of g. This would entail finding

$$g^* = \arg\min_{g \in \mathcal{G}} J(g) = \arg\min_{g \in \mathcal{G}} \sum_{i,j} |d_{ij} - v(t_i, \xi_j; g)|^2, \qquad (14.1)$$

where $\mathcal{G}$ is a family of admissible growth rates for a given population of mosquitofish. However for such problems tracking of individuals is usually impossible. Thus we turn to methods where one can use aggregate data.

14.1 Populations with Aggregate Data, Uncertainty, and PBM

14.1.1 Type I: Individual Dynamics/Aggregate Data Inverse Problems

Recall from Example 5 that the expected size density $u(t, \xi; P) = \mathcal{E}[v(t, \xi; \cdot)|P]$ is described by a general probability distribution P and is given by

$$u(t, \xi; P) = \int_{\mathcal{G}} v(t, \xi; g) dP(g),$$

where the density $v(t, \xi; g)$ satisfies, for a given g, the Sinko-Streifer system (1.22)-(1.23).

In these problems, even though one has individual dynamics (the $v(t, \xi; g)$ for a given cohort of fish with growth rate g), one has only *aggregate* or population level longitudinal data available. This is common in marine, insect, etc., *catch and release* experiments [BK] where one samples at different times from the same population but cannot be guaranteed of observing the same set of individuals at each sample time. This type of data is also typical in experiments where the organism or population member being studied is sacrificed in the process of making a single observation (e.g., certain physiologically based pharmacokinetic (PBPK) modeling [BPo, Po] and whole organism transport models [BK]). In these cases one may still have dynamic (i.e., time course) models for individuals as in the mosquitofish, but no individual data is available.

Since we must use aggregate population data d_{ij} to estimate P itself in the corresponding typical inverse problems, we are therefore required to understand the qualitative properties (continuity, sensitivity, etc.) of $u(t, \xi; P)$ as a function of P. The data for parameter estimation problems are d_{ij}, which observations for $u(t_i, \xi_j; P)$. The corresponding OLS inverse problem consists of minimizing

$$J(P) = \sum_{ij} |d_{ij} - u(t_i, \xi_j; P)|^2 \tag{14.2}$$

over $P \in \mathcal{P}(\mathcal{G})$, the set of probability distributions on $\mathcal{G}$, or over some suitably chosen subset of $\mathcal{P}(\mathcal{G})$.

14.1.2   Type II: Aggregate Dynamics/Aggregate Data Inverse Problems

The second class of problems which we call *Type II* problems involves dynamics which depend explicitly on the probability distribution P itself. In this case one only has dynamics (*aggregate dynamics*) for the expected value of the population state variable. No dynamics are available for individual trajectories $x(t, q)$ for a given $q \in Q$. The electromagnetic example of Example 4 of Section 1.6 is precisely this situation. Such problems also arise in viscoelasticity as well as biology (the HIV cellular models of Banks, Bortz and Holte [BBH]) — see also [BBPP, BG1, BG2, BPi].

To illustrate this second type of inverse problem we return to the electromagnetics example of Example 4 which entails polarization of inhomogeneous dielectric materials. In this case, individual (particle or molecular) dynamics are not available. Instead, the dynamics (1.17) (in second-order form for either E or H) themselves depend on a probability measure F, e.g.,

$$\frac{\partial^2 u}{\partial t^2} + \frac{\partial^2 u}{\partial x^2} = f(u, F),$$

where for example the electromagnetic field $u = E$ depends on summing with a probability distribution the effects of multiple mechanisms of polarizations across all particles in the material. To be a little more specific, we consider multiple relaxation mechanisms in a dielectric material.

To describe the behavior of the electric polarization P, we begin with the general formulation of Example 4 of Section 1.6 and Chapter 2 of [BBL] by employing a polarization kernel g in the convolution expression

$$P(t, z) = \int_0^t g(t - s, z; \tau) E(s, z) ds. \tag{14.3}$$

As explained in Section 1.6, this general formulation includes as special cases (at the particle level) the well known orientational or Debye polarization model, the electronic or Lorentz polarization model, and linear combinations thereof, as well as other higher order models depending on relaxation parameters τ. However, use of these kernels presupposes that the material may be sufficiently defined by a single relaxation parameter τ, which is generally not the case. In order to account for multiple relaxation parameters in the polarization mechanisms, we allow for a distribution of relaxation parameters which is conveniently described in terms of a probability measure F on a set of possible relaxation values $\mathcal{T}$. Thus, we define our polarization model in terms of a convolution operator

$$P(t, z) = \int_0^t G(t - s, z; F) E(s, z) ds$$

where G is determined by various polarization mechanisms each described by a different parameter τ, and therefore is given by

$$G(t, z; F) = \int_{\mathcal{T}} g(t, z; \tau) dF(\tau)$$

where $\mathcal{T} \subset [\tau_1, \tau_2]$.

To obtain the macroscopic polarization, we sum over all the parameters. We cannot separate dynamics for the electric and magnetic fields to obtain individual or particle dynamics. Therefore we have (1.17) in Section 1.6 as an example where the dynamics for the E and H fields depend explicitly on the probability measure $F \in \mathcal{P}(\mathcal{T})$.

For inverse problems, data is given in terms of field measurements d_{ij} for $E(t_i, z_j; F)$. In ordinary least square problems, we have

$$J(F) = \sum_{ij} |d_{ij} - E(t_i, z_j; F)|^2, \qquad (14.4)$$

to be minimized over $F \in \mathcal{P}(\mathcal{T})$ (or over some subset of $\mathcal{P}(\mathcal{T})$).We note that while (14.4) and (14.2) may appear similar, this appearance is somewhat misleading. In (14.2) we sum with the probability measure over a family of "individual" dynamics, while in (14.4), the equation for the observable depends explicitly on the probability distribution of interest.

Thus, to carry out analysis of the minimization problems using (14.2) or (14.4), we must have a concept of regularity (in the topological or metric sense) in probability spaces.

More fundamentally for Type II systems, the "well-posedness of systems," including existence and continuous dependence of solutions on "parameters," depends on the concept of $E(t, z; F_1)$ and $E(t, z; F_2)$ being close whenever F_1 and F_2 are close, i.e., continuity of $F \mapsto E(t, z; F)$.

14.2 A Prohorov Metric Framework for Inverse Problems

We consider formulation and approximation methods (we shall call this the **Prohorov Metric Framework (PMF)**) for estimation or inverse problems where the quantity of interest is a probability distribution P defined on a metric space Q. To simply illustrate ideas we assume we have a Type I problem with a parameter ($q \in Q$, Q a metric space) dependent system with model responses or system dynamics $y(t, q)$ describing the population of interest. For data or observations, we are given a set of data values $\{d_l\}$ for the expected values

$$\mathcal{E}[y_l(q)|P] = \int_Q y_l(q)dP(q)$$

of the model solutions $y_l(q) = y(t_l, q)$ with respect to an unknown probability distribution P describing the distribution of parameters q over a population. We wish to use this data to choose from a given family $\mathcal{P}(Q)$ the distribution P^* that gives a best fit of the underlying model to data. In our presentation here we will, to simplify discussions, choose $\mathcal{P}(Q)$ as the set of probability distributions on a metric space Q although one could use the same formulation if one restricts the problem to a properly chosen subset of $\mathcal{P}(Q)$.

To describe these problems and methodology, we formulate ordinary least squares (OLS) problems; this is not essential as one could equally well use a weighted least squares (WLS), a generalized least squares (GLS), a maximum likelihood estimator (MLE), etc., approach [BDSS]. In a fundamental OLS problem, one seeks to minimize

$$J(P) = \sum_l |\mathcal{E}[y_l(q)|P] - d_l|^2 \tag{14.5}$$

over $P \in \mathcal{P}(Q)$.

Even for simple dynamics with evaluations $\{y_l(q)\}$, this is an infinite dimensional optimization problem, so that one needs approximations that lead to computationally tractable schemes. That is, it is useful to formulate methods to yield finite dimensional sets $\mathcal{P}^M(Q)$ over which to minimize $J(P)$. Of course, we wish to choose these methods so that "$\mathcal{P}^M(Q) \approx \mathcal{P}(Q)$" in some sense. In this case we shall use the *Prohorov metric* [BBPP, BBi] of weak* convergence of measures to obtain the desired approximation results. This metric, along with others, will be defined and discussed in some detail in the next section.

A general theoretical framework developed during the past several decades is outlined in [BBPP] with specific results on the approximations we use here given in [BBi, BPi]. Briefly, ideas for the underlying theory are as follows:

1. One argues *continuity of* $P \to J(P)$ on $\mathcal{P}(Q)$ with the Prohorov metric;

2. If the metric space Q is compact then $\mathcal{P}(Q)$ is a *compact metric space* when taken with Prohorov metric;

3. Approximation families $\mathcal{P}^M(Q)$ are chosen so that elements $P^M \in \mathcal{P}^M(Q)$ can be found to approximate elements $P \in \mathcal{P}(Q)$ in the Prohorov metric;

4. *Well-posedness* (existence, continuous dependence of estimates on data, etc.) is obtained along with *feasible computational methods.*

The data $\{d_l\}$ and dynamics available (individual or aggregate, either of which, in general, will involve longitudinal or time evolution data) determines the nature (Type I or Type II) of the inverse problem. While the approximations we discuss below are applicable to both types of problems, we shall illustrate the computational results in the context of *size-structured marine populations (mosquitofish, shrimp)* where the inverse problems are of Type I. Finally, we note that in the problems considered here, one *cannot sample directly from the probability distribution* being estimated and this is somewhat different from the usual case treated in some of the statistical literature, e.g., see [Wahba1, Wahba2] and the references cited therein.

Before continuing our discussions, we first digress briefly to discuss possible metrics on sets or spaces $\mathcal{P}(Q)$ of probability measures or distributions on complete metric spaces Q.

14.3   Metrics on Probability Spaces

We consider $\mathcal{P}(Q)$ as the space of probability measures or distributions on a separable complete metric space Q. Let P_1 and P_2 be probability distributions. If working on the real line $(-\infty, \infty)$ or $(0, \infty)$ we will sometimes not distinguish between the probability measure P and its cumulative distribution function F.

We need to understand the concept of a metric or distance $\rho(P_1, P_2)$ between P_1 and P_2. Moreover, in view of the discussions in the previous section, we will be interested in compactness with respect to these metrics on $\mathcal{P}(Q)$. We recall (see Section 4.3, Theorem 4.5) that $\text{rba}(Q) \cong C_B^*(Q)$ and hence one can consider $\mathcal{P}(Q) \subset \text{rba}(Q)$ as a subset of $C_B^*(Q)$ and consider the corresponding weak* topology on $\mathcal{P}(Q)$. There are several metrics that metrize (i.e., generate an equivalent topology for) the weak* topology on $\mathcal{P}(Q) \subset C_B^*(Q)$ including

i) The Levy metric, denoted here by $\rho_L(P_1, P_2)$,

ii) The Prohorov metric, denoted by $\rho(P_1, P_2) = \rho_{PR}(P_1, P_2)$,

iii) The bounded Lipschitz metric, denoted by $\rho_{BL}(P_1, P_2)$.

Others that do not metrize the weak* topology on $\mathcal{P}(Q)$ include

iv) The total variation metric, denoted by $\rho_{TV}(P_1, P_2)$,

v) The Kolmogorov metric, denoted by $\rho_K(P_1, P_2)$.

We discuss each of these here; for relevant material, see [B, H, P].

14.3.1 The Prohorov Metric

Previous developments in probability theory provide helpful results in the pursuit of a possible complete computational methodology (i.e., a computationally tractable approximation framework). One of the most important tools in probability theory is the *Prohorov metric*, which we will now formally define. Let Q be a complete separable metric space with metric d. Given a closed subset A of Q, define the ϵ-neighborhood of A as

$$
\begin{aligned}
A^\epsilon &= \{q \in Q : d(\hat{q}, q) \le \epsilon \text{ for some } \hat{q} \in A\} \\
&= \{q \in Q : \inf_{\hat{q} \in A} d(\hat{q}, q) \le \epsilon\}.
\end{aligned}
$$

We define the Prohorov metric $\rho = \rho_{PR} : \mathcal{P}(Q) \times \mathcal{P}(Q) \to \mathbb{R}^+$ by

$$
\rho(P_1, P_2) \equiv \inf\{\epsilon > 0 : P_1(A) \le P_2(A^\epsilon) + \epsilon, \ A \text{ closed } \subset Q\}.
$$

This can be shown to be a metric on $\mathcal{P}(Q)$ and has many properties including

(a.) If Q is a complete separable metric space, then $(\mathcal{P}(Q), \rho)$ is a complete separable metric space;

(b.) If Q is compact, then $(\mathcal{P}(Q), \rho)$ is a compact metric space.

Note that the definition of ρ is not intuitive. For example, we do not necessarily know what $P_k \to P$ in ρ means. We have the following important and useful characterizations [B]. Given $P_k, P \in \mathcal{P}(Q)$, the following convergence statements are equivalent:

1. $\rho(P_k, P) \to 0$;

2. $\int_Q f dP_k(q) \to \int_Q f dP(q)$ for all bounded, continuous $f : Q \to \mathbb{R}^1$;

3. $P_k[A] \to P[A]$ for all Borel sets $A \subset Q$ with $P[\partial A] = 0$.

Thus, we immediately obtain the following results:

- Convergence in the ρ metric is equivalent to convergence in distribution or so-called "weak" convergence of measures encountered in probability theory.

- Let $C_B^*(Q)$ denote the topological dual of $C_B(Q)$, where $C_B(Q)$ is the usual space of bounded continuous functions on Q with the supremum norm (see Section 4.3). If we view $\mathcal{P}(Q) \subset C_B^*(Q)$, convergence in the ρ topology is equivalent to weak* convergence in $\mathcal{P}(Q)$. Note the misnomer "weak convergence of measures" used by probabilists is actually weak* convergence in the functional analysis sense defined here and in almost all standard texts.

More importantly,

$$\rho(P_k, P) \to 0 \text{ is equivalent to } \int_Q y(t_i; q) dP_k(q) \to \int_Q y(t_i; q) dP(q),$$

for any continuous functions $q \to y(t_i; q)$. Thus $P_k \to P$ in the ρ metric is equivalent to

$$\mathcal{E}[y(t_i; q)|P_k] \to \mathcal{E}[y(t_i; q)|P]$$

or "convergence in expectation." This yields that

$$P \to J(P) = \sum_{i=1}^{n} |\mathcal{E}[y(t_i; q)|P] - \hat{d}_i|^2$$

is continuous in the ρ topology. Continuity of $P \to J(P)$ allows us to assert the existence of a solution to $\min J(P)$ over $P \in \mathcal{P}(Q)$ defined in (14.5) whenever Q, and hence $\mathcal{P}(Q)$, is compact.

These considerations allow us to state the following theorem [B, p. 238].

Theorem 14.1 *The Prohorov metric metrizes the weak* topology of* $\mathcal{P}(Q)$.

Now we may consider whether there are other (equivalent) metrics that metrize the weak* topology on $\mathcal{P}(Q)$. But first we point out an application in statistics where the need for metrics on distributions arises.

14.3.2   Robust Statistics

Prohorov (also sometimes found as Prokhorov in translations) was interested in finding a useful metric for the frequently used "convergence in distribution" or "convergence in measure" in probability theory. Later theoretical statisticians became interested in (and concerned

with) the underlying foundations for statistical testing and in particular inference procedures [H]. Specifically, attention was given to the
"stability" of assumptions under which statistical asymptotic theories
may be valid. This led to "robustness" (insensitivity to small deviations in underlying assumptions) or *Robust Statistics*. These ideas focus
on "distributional robustness" as embodied in insensitivity to "small"
deviations in distributions (often from the normal or Gaussian distribution assumptions found in many formulations) and also how lack of
this robustness might affect the validity of statistical tests. This led to
a renewed interest in the use of distributional metrics such as that of
Prohorov. As formulated by Prohorov this led to attempts to find a
metric for the "weak" topology on $\mathcal{P}(Q)$ defined as the weakest topology on $\mathcal{P}(Q)$ such that for every bounded continuous ψ $(\psi \in C_B(Q))$
the map

$$P \to \int_Q \psi dP$$

is continuous, i.e., consider $\mathcal{P}(Q) \subset C_B^*(Q)$.

To describe the various metrics that have found use by statisticians
and probabilitists, it is desirable to recall the concept of a *Polish space*
which is defined as a complete, separable, metrizable topological space.
We turn to a brief summary of other metrics of interest.

14.3.3 The Levy Metric

The Levy metric is defined in the special case for $Q = \mathbb{R}^1$ and is the
Prohorov metric restricted to this case, where one usually does not
distinguish between the probability measure or distribution P_1 and its
cumulative distribution function (cdf) F_1. The metric is given by

$$\rho_L(P_1, P_2) = \inf\{\epsilon| \text{ for all } x, P_1(x - \epsilon) - \epsilon \le P_2(x) \le P_1(x + \epsilon) + \epsilon\}.$$

One can argue that this is a symmetric function and indeed defines a
metric.

Remarks

- $\sqrt{2}\rho_L(P_1, P_2)$ is the maximum distance between graphs of P_1, P_2
  measured along the 45° direction. It follows from results for the
  Prohorov metric that the Levy metric metrizes the weak* topology of $\mathcal{P}(\mathbb{R}^1)$.

- The Prohorov metric is more difficult to visualize but is applicable when Q is any complete separable metric space (Polish space), not just the real line (Levy case). For example, when Q is a function space such as growth rates or mortality rates in Sinko-Streifer (mosquito fish problem), the Prohorov metric is applicable whereas the Levy metric is not.

We have already noted that both the Levy and Prohorov metrics metrize the weak* topology on $\mathcal{P}(Q)$, the former only when $Q = \mathbb{R}^1$. Moreover, as noted above, we can argue that for Q a complete separable metric space, then $(\mathcal{P}(Q), \rho_{PR})$ is a complete separable metric space.

Recall separability implies the existence of a subset Q_0 that is a countable dense subset of Q. One can then obtain a countable dense subset of $\mathcal{P}$ by defining (see also Theorem 14.3 below)

$$
\begin{aligned}
\mathcal{P}_0 &= \{\text{measures with finite support in } Q_0 \text{ with rational masses}\} \\
&= \{P = \sum_{\text{finite}} p_i \delta_{q_i} | p_i \text{ rational}, \{q_i\} \subset Q_0, \Sigma p_i = 1\}
\end{aligned}
$$

14.3.4 The Bounded Lipschitz Metric

Assume without loss of generality that distance function d on Q is bounded by 1. (If necessary, one can replace the metric by an equivalent metric $\tilde{d}(q_1, q_2) = \frac{d(q_1,q_2)}{1+d(q_1,q_2)}$.) Then define

$$
\rho_{BL}(P_1, P_2) = \sup_{\psi \in \Psi} \left| \int_Q \psi dP_1 - \int_Q \psi dP_2 \right|
$$

$$
\text{where } \Psi = \{\psi \in C_B(Q) : |\psi(q_1) - \psi(q_2)| \le d(q_1, q_2)\}
$$

A fundamental result [H, p.33] of practical importance is

Theorem 14.2 *For all* $P_1, P_2 \in \mathcal{P}(Q)$,

$$
\rho_{PR}(P_1, P_2)^2 \le \rho_{BL}(P_1, P_2) \le 2\rho_{PR}(P_1, P_2).
$$

Thus, ρ_{PR} and ρ_{BL} define the same topology and hence ρ_{BL} also metrizes the weak* topology on $\mathcal{P}(Q)$. This knowledge about ρ_{BL} versus ρ_{PR} can be quite useful. We have that

$$
\begin{aligned}
\Delta_k &\equiv \left| \int_Q \psi(q) dP_k(q) - \int_Q \psi(q) dP(q) \right| \\
&\le \rho_{BL}(P_k, P) \\
&\le 2\rho_{PR}(P_k, P).
\end{aligned}
$$

Therefore, if we know $P_k \to P$ in ρ_{PR}, it may be useful in direct estimates for error rates and convergence analyses with the Prohorov metric since $\Delta_k \to 0$ bounds the square of the Prohorov metric if $\psi \in C_B(Q)$.

14.3.5 Other Metrics

Other metrics of interest can be found in the literature [H]. These include

- Total variation metric:

$$\rho_{TV}(P_1, P_2) = \sup_{A \in \mathcal{B}} |P_1(A) - P_2(A)|,$$

  where $\mathcal{B}$ are the Borel subsets of Q, and

- Kolmogorov metric: $(Q = \mathbb{R}^1)$

$$\rho_K(P_1, P_2) = \sup_{x \in \mathbb{R}^1} |P_1(x) - P_2(x)|.$$

These metrics do not metrize the weak* topology, but do satisfy

$$\rho_L \leq \rho_{PR} \leq \rho_{TV}$$
$$\rho_L \leq \rho_K \leq \rho_{TV}.$$

14.4 Example 5: The Growth Rate Distribution Model and Inverse Problem in Marine Populations

Continuing with our motivating application (one could also use the electromagnetics models of Example 4, e.g., see [BG1, BG2]), we find that the problem entails estimation of growth rate distributions for size-structured mosquitofish populations. Mosquitofish are used in place of pesticides to control mosquito populations in rice fields. Marine biologists desire to correctly predict growth and decline of the mosquitofish populations in order to determine the optimal densities of mosquitofish to use to control mosquito populations. A mathematical model that accurately describes the mosquitofish population would be beneficial in this application, as well as in other problems involving population dynamics and age/size-structured data.

Based on data (Figure 14.1) collected from rice fields, a qualitatively reasonable mathematical model must predict two key features that are exhibited in the data: *dispersion* and *bifurcation* (i.e., a unimodal density becomes a bimodal density) of the population density over time [BBKW, BF, BFPZ]. As mentioned above in Section 1.7, the Sinko-Streifer model does not exhibit either of these under reasonable biological assumptions. However, the *Growth Rate Distribution (GRD) model*, developed in [BBKW] and [BF], captures both of these features in its solutions.

The model is a modification of the *Sinko-Streifer model* (used for modeling age/size-structured populations) which allows individuals to have different characteristics or behaviors with respect to growth.

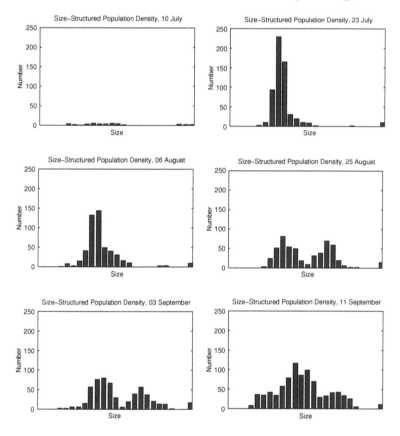

Figure 14.1: Mosquitofish data (number vs. size) for the days July 10, 23, August 6, 25, and September 3, 11.

Recall that the *Sinko-Streifer model (SS)* for size-structured mosquitofish populations is given by equation (1.22), which for convenience we repeat as

$$\frac{\partial v}{\partial t} + \frac{\partial}{\partial \xi}(gv) = -\mu v, \quad \xi_0 < \xi < \xi_1, \quad t > 0 \qquad (14.6)$$

$$v(0, \xi) = \Phi(\xi)$$

$$g(t, \xi_0)v(t, \xi_0) = \int_{\xi_0}^{\xi_1} K(t, s)v(t, s)ds$$

$$g(t, \xi_1) = 0.$$

Recall also that $v(t, \xi)$ represents size or population density (given in numbers per unit length), where t represents time and ξ represents length of mosquitofish. The growth rate of an individual mosquitofish is given by $g(t, \xi)$ so that

$$\frac{d\xi}{dt} = g(t, \xi) \qquad (14.7)$$

for each individual (all mosquitofish of a given size have the same growth rate).

As we have noted, the SS model cannot be used in its usual form to model the mosquitofish population because it *does not predict dispersion or bifurcation* of the population in time under biologically reasonable assumptions [BBKW, BF]. But by modifying the SS model so that the individual growth rates of the mosquitofish *vary across the population* (instead of being the same for all individuals in the population), one obtains a model, the GRD model, which does in fact exhibit both dispersal in time and development of a bimodal density from a unimodal density (see [BF, BFPZ]).

As noted earlier in discussions of Example 5, in the GRD model, the population density $u(t, \xi; P)$ is actually given by

$$u(t, \xi; P) = \int_{\mathcal{G}} v(t, \xi; g)dP(g), \qquad (14.8)$$

where $\mathcal{G}$ is a collection of admissible growth rates, P is a probability measure on $\mathcal{G}$, and $v(t, \xi; g)$ is the solution of the (SS) equation (14.6) with g. This model assumes that the population is made up of *collections of subpopulations*—individuals in the same subpopulation have the same growth rate. As demonstrated in [BF], solutions to the GRD model exhibit both dispersion and bifurcation of the population density in time.

To illustrate ideas further, we assume that the *admissible growth rates g* have the form

$$g(\xi; b, \gamma) = b(\gamma - \xi)$$

for $\xi_0 \le \xi \le \gamma$ and zero otherwise, where b is the *intrinsic growth rate* of the mosquitofish and $\gamma = \xi_1$ is the maximum size. This choice is based on work in [BBKW], where other ideas of properties related to the growth rates varying among the mosquitofish are discussed.

Under the assumption of varying intrinsic growth rates and maximum sizes, we assume that b and γ are *random variables* taking values in the compact sets B and Γ, respectively. A reasonable assumption is that both are *bounded closed intervals*. Thus we take

$$\mathcal{G} = \{g(\cdot; b, \gamma) | b \in B, \gamma \in \Gamma\}$$

so that $\mathcal{G}$ is also compact in, for example, $C[\xi_0, \xi_1]$ where $\xi_1 = \max(\Gamma)$. Then $\mathcal{P}(G)$ is compact in the Prohorov metric and we are in the framework outlined above. To further simplify our example, one may choose growth rate functions parameterized by the intrinsic growth rate b with $\gamma = 1$, leading to a one parameter family of varying growth rates g among the individuals in the population. We may also assume that $\mu = 0$ and $K = 0$ in order to focus on only the distribution of growth rates; however, distributions could just as well be placed on μ and K.

Next we introduce two different approaches that can be used in inverse problems for estimation of distributions of growth rates in the mosquitofish example.

The *first approach*, which has been discussed and used in [BF] and [BFPZ], involves the use of finite convex combinations of *Dirac delta distributions*. We assume that probability distributions $\mathcal{P}^M$ placed on growth rates are *discrete* corresponding to a collection $\mathcal{G}^M$ with the form $\mathcal{G}^M = \{g_k\}_{k=1}^{M}$ where $g_k(\xi) = b_k(1 - \xi)$, for $k = 1, \ldots, M$, where the $\{b_k\}$ are a *discretization* of B. For each subpopulation k with growth rate g_k, there is a corresponding probability p_k that an individual is in subpopulation k. The population density $u(t, \xi; P)$ in (14.8) is then approximated by

$$u(t, \xi; \{p_k\}) = \sum_{k=1}^{M} v(t, \xi; g_k) p_k,$$

where $v(t, \xi; g_k)$ is the subpopulation density from equation (14.6) with growth rate g_k. We denote this *delta function approximation method*

as *DEL(M)*, where M is number of elements used in this approximation. Theoretical results for these approximations are embodied in the
following result [BBi]:

Let Q be a complete, separable metric space, $\mathcal{B}$ the class of all Borel
subsets of Q and $\mathcal{P}(Q)$ the space of probability measures on $(Q, \mathcal{B})$.
Let $Q_0 = \{q_j\}_{j=1}^\infty$ be a countable, dense subset of Q. Then the set of
$P \in \mathcal{P}(Q)$ such that P has finite support in Q_0 and rational masses is
dense in $\mathcal{P}(Q)$ in the Prohorov metric.

Theorem 14.3 *That is,*

$$\mathcal{P}_0(Q) \equiv \left\{ P \in \mathcal{P}(Q) : P = \sum_{j=1}^k p_j \Delta_{q_j}, k \in \mathcal{N}^+, \right.$$

$$\left. q_j \in Q_0, \ p_j \ rational, \sum_{j=1}^k p_j = 1 \right\}$$

*is dense in $\mathcal{P}(Q)$ taken with the Prohorov metric, where Δ_{q_j} is the
Dirac measure with atom at q_j.*

It is rather easy to use the ideas and results associated with this
theorem to develop computationally efficient schemes. Given $Q_d = \bigcup_{M=1}^\infty Q_M$ with $Q_M = \{q_j^M\}_{j=1}^M$ (a "partition" of Q) chosen so that Q_d
is dense in Q, define

$$\mathcal{P}^M(Q) = \left\{ P \in \mathcal{P}(Q) : P = \sum_{j=1}^M p_j \Delta_{q_j^M}, q_j^M \in Q_M, \ p_j \ rational, \sum_{j=1}^M p_j = 1 \right\}.$$

$$(14.9)$$

Then we find

Lemma 14.1 *Let $\mathcal{P}(Q)$ be the metric space of probability measures on
Q taken with the Prohorov metric ρ and let $\mathcal{P}^M(Q)$ be defined as in
(14.9). Then we have*

(i) $\mathcal{P}^M(Q)$ is a compact subset of $\mathcal{P}(Q)$ in the ρ metric,

(ii) $\mathcal{P}^M(Q) \subset \mathcal{P}^{M+1}(Q)$ whenever Q_{M+1} is a refinement of Q_M,

*(iii) "$\mathcal{P}^M(Q) \to \mathcal{P}(Q)$" in the ρ topology; that is, for M sufficiently
large, elements in $\mathcal{P}(Q)$ can be approximated in the ρ metric by
elements of $\mathcal{P}^M(Q)$.*

While it has been shown [BBi, BD] that DEL(M) provides a reasonable approximation to (14.8), another (and perhaps better?) approach might involve techniques that will provide a smoother approximation of (14.8) in the case of continuous probability distributions on the growth rates. Thus, as a *second approach*, we might choose to use an approximation scheme based on piecewise linear splines. In this case we assume that P is a continuous (actually absolutely continuous) probability distribution on the intrinsic growth rates. We approximate the density $P' = \frac{dP}{db} = p(b)$ using piecewise linear splines, which leads to the following approximation for $u(t, \xi; P)$ in (14.8):

$$u(t, \xi; \{a_k\}) \approx \sum_{k=1}^{M} a_k \int_B v(t, \xi; g(\xi; b)) l_k(b) db,$$

where $g(\xi; b) = b(1 - \xi)$, $p_k(b) = a_k l_k(b)$ is the probability density for an individual in subpopulation k and the l_k are piecewise linear spline functions. This spline based approximation method is denoted by SPL(M,N), where M is the number of basis elements used to approximate the growth rate probability distribution and N is the number of quadrature nodes used to approximate the integral in the formula above. A theoretical approximation result [BPi] provides a rigorous foundation for this approach.

Theorem 14.4 *Let $\mathcal{F}$ be a weakly compact subset of $L_2(Q)$, Q compact and let $\mathcal{P}_\mathcal{F}(Q) \equiv \{P \in \mathcal{P}(Q) : P' = p, \ p \in \mathcal{F}\}$. Then $\mathcal{P}_\mathcal{F}(Q)$ is compact in $\mathcal{P}(Q)$ in the Prohorov metric. Moreover, if we define $\{\ell_j^M\}$ to be the linear splines on Q corresponding to the partition Q_M, where $\bigcup_M Q_M$ is dense Q, define*

$$\mathcal{P}^M \equiv \{p^M : p^M = \sum_j b_j^M \ell_j^M, b_j^M \ \text{rational}\}$$

and if

$$\mathcal{P}_{\mathcal{F}M} \equiv \{P_M \in \mathcal{P}(Q) : P'_M = p^M, p^M \in \mathcal{P}^M\},$$

we have $\bigcup_M \mathcal{P}_{\mathcal{F}M}$ is dense in $\mathcal{P}_\mathcal{F}(Q)$ taken with the Prohorov metric.

A computational study comparing the relative strengths and weaknesses of these two classes of approximation schemes in the context of inverse problems for estimating probability measures is given in [BD].

One can use the approximation methods DEL(M) and SPL(M,N) in the inverse problem for the estimation of the growth rate distributions. The least squares inverse problem to be solved is

$$\min_{P \in \mathcal{P}^M(\mathcal{G})} J(P), \qquad (14.10)$$

where

$$
\begin{aligned}
J(P) &= \sum_{i,j} (u(t_i, \xi_j; P) - \hat{u}_{ij})^2 \qquad (14.11) \\
&= \sum_{i,j} (u(t_i, \xi_j; P))^2 - 2u(t_i, \xi_j; P)\hat{u}_{ij} + (\hat{u}_{ij})^2,
\end{aligned}
$$

in which $\{\hat{u}_{ij}\}$ is the data and $\mathcal{P}^M(\mathcal{G})$ is the finite dimensional approximation to $\mathcal{P}(\mathcal{G})$. When using DEL(M), the finite dimensional approximation $\mathcal{P}^M(\mathcal{G})$ to the probability measure space $\mathcal{P}(\mathcal{G})$ is given by

$$\mathcal{P}^M(\mathcal{G}) = \left\{ P \in \mathcal{P}(\mathcal{G}) \mid P = \sum_k p_k \Delta_{g_k}, \ \sum_k p_k = 1 \right\},$$

where Δ_{g_k} is the delta distribution with an atom at g_k. When using SPL(M,N), the finite dimensional approximation $\mathcal{P}^M(\mathcal{G})$ is given by

$$\mathcal{P}^M(\mathcal{G}) = \left\{ P \in \mathcal{P}(\mathcal{G}) \mid P' = \sum_k a_k l_k(b), \ \sum_k a_k \int_B l_k(b) db = 1 \right\}.$$

Furthermore, we note that these least squares inverse problems (14.10)-(14.11) become quadratic programming problems [BD, BF, BFPZ], for which highly efficient computational routines exist. Letting **p** be the vector that contains p_k, $1 \le k \le M$, when using DEL(M) or a_k, $1 \le k \le M$, when using SPL(M,N), we let **A** be the matrix with entries given by

$$A_{km} = \sum_{i,j} v(t_i, \xi_j; g_k) v(t_i, \xi_j; g_m),$$

let **b** the vector with entries given by

$$b_k = -\sum_{i,j} \hat{u}_{ij} v(t_i, \xi_j; g_k),$$

and

$$c = \sum_{i,j} (\hat{u}_{ij})^2,$$

where $1 \le k, m \le M$. In the place of (14.11), we now minimize

$$F(\mathbf{p}) \equiv \mathbf{p}^T \mathbf{A} \mathbf{p} + 2\mathbf{p}^T \mathbf{b} + c \qquad (14.12)$$

over $\mathcal{P}^M(\mathcal{G})$. We note when using DEL(M) we also have to include the constraint

$$\sum_k p_k = 1,$$

while when using SPL(M,N) we have to include the constraint

$$\sum_k a_k \int_B l_k(b) db = 1.$$

However, in both cases, one is able to include these constraints along with nonnegativity constraints on the $\{p_k\}$ and $\{a_k\}$ in the programming of these two inverse problems. Again, the reader may consult [BD] for comparative results.

Other Size-Structured Population Models

The Sinko-Streifer (SS) model [SS] and its variations have been widely used to describe numerous age and size-structured populations (see [BBDS1, BBDS2, BBKW, BF, BFPZ, BA, Kot, MetzD] for example).

One recent example involves the shrimp growth models developed and employed in [BaBoHetalShrimp, BDEHAD]. Dispersion in size in experimental data for early growth of shrimp has been observed in different raceways at the Shrimp Mariculture Research Facility, Texas Agricultural Experiment Station in Corpus Christi, TX. Initial sizes were very similar but variability is observed in aggregate type longitudinal data. A reasonable model for these populations must account for variability in size distribution data which perhaps is a result of variability in individual growth rates across populations.

In summary, the GRD model (14.8) represents one approach to accounting for variability in growth rates by imposing a probability distribution on the growth rates in the SS model (1.22) or (14.6). Individuals in the population grow according to a *deterministic growth model* (14.7), but different individuals in the population may have different parameter dependent growth rates in the GRD model. The population is assumed to consist of *subpopulations* with individuals in the same subpopulation having the same growth rate. The growth uncertainty of individuals in the population is the result of *variability in growth rates among the subpopulations.* This modeling approach, which entails a stationary probabilistic structure on a family of deterministic dynamic systems, may be most appropriate when the growth of individuals is assumed to be the result of *genetic variability* or varying levels of *chronic disease or infection.*

Another approach to models with variability in size involves the use of stochastic differential equations where the growth process itself is assumed stochastic. The resulting models lead to Fokker-Planck or Forward Kolmogorov partial differential equations for the associated probability densities in time and size. Recent efforts [GRD-FP, GRD-FP2, BHu] have discussed the relationship and certain equivalences between the two distinct approaches.

15 The Prohorov Metric in Optimization and Optimal Design Problems

Before proceeding to discuss functional analytic formulations of control problems as promised earlier, in a brief digression we return to problems with uncertainty in which the Prohorov metric of weak* convergence of Chapter 14 plays a significant role. In particular we mention some older areas (relaxed controls, hysteresis in materials) along with some more recent applications in two person min-max games and optimal design.

15.1 Two Player Min-Max Games with Uncertainty

We consider electromagnetic evasion-interrogation games wherein the evader can use ferroelectric material coatings to attempt to avoid detection while the interrogator can manipulate the interrogating frequencies (wave numbers) and angles of incidence of the interrogating inputs to enhance detection and identification. The resulting problems are formulated as two player games in which one player wishes to minimize the reflected signal while the other wishes to maximize it. Simple deterministic strategies are easily defeated and hence the players must introduce uncertainty to disguise their intentions and confuse their opponent. Mathematically, the resulting game is carried out over spaces of probability measures which in many cases are appropriately metrized using the Prohorov metric. Details of the results discussed in this section can be found in [BGIT].

In [BIKT] the authors demonstrated that it is possible to design ferroelectric materials with appropriate dielectric permittivity and magnetic permeability to significantly attenuate reflections of electromagnetic interrogation signals from highly conductive targets such as airfoils and missiles. This was done under assumptions that the interrogating input signal is uniformly likely to come from a sector of interrogating angles $\alpha \in [\alpha_0, \alpha_1]$ ($\alpha = \frac{\pi}{2} - \phi$ where ϕ is the angle of incidence of the input signal) but that the evader has knowledge of the interrogator's input frequency or frequencies (denoted in the presentations below as the *interrogator design frequencies* I_D). These results were further sharpened and illustrated in [BIT] where a series of different material designs were considered to minimize over a given set of input design frequencies I_D the maximum reflected field from input signals.

In addition, a second critical finding was obtained in that it was shown that if the evader employed a simple counter interrogation design based on a fixed set (assumed known) of interrogating frequencies I_D, then by a rather simple counter-counter interrogation strategy (use an interrogating frequency little more than 10% different from the assumed design frequencies), the interrogator can easily defeat the evader's material coatings counter interrogation strategy to obtain strong reflected signals.

From the combined results of [BIKT, BIT] it is thus rather easily concluded that the evader and the interrogator must each try to confuse the other by introducing significant *uncertainty* in their design and interrogating strategies, respectively. This concept, which we refer to as *mixed strategies* in recognition of previous contributions to the literature on games (von Neumann's *finite mixed strategies* to be explained below), leads to two player non-cooperative games with probabilistic strategy formulations. These can be mathematically formulated as two sided optimization problems over spaces of probability measures, i.e., min-max games over sets of probability measures.

15.1.1   Problem Formulation

We consider electromagnetic interrogation of objects in the context of min-max evader-interrogator games where each player has uncertain information about the adversary's capabilities. The min-max cost functional is based on reflected fields from an object such as an airfoil or missile and can be computed in one of several ways [BIKT, BIT].

The simplest computational method employs the reflection coefficient based on a simple planar geometry (e.g., see Figure 15.1) using Fresnel's formula for a perfectly conducting half plane which has a coating layer of thickness d with dielectric permittivity ϵ and magnetic permeability μ. A normally incident ($\phi = 0$) electromagnetic wave with the frequency f is assumed to impinge the half plane. Then the corresponding wavelength λ in air is $\lambda = c/f$, where the speed of light is $c = 0.3 \times 10^9$. The reflection coefficient R for the wave is given by

$$R = \frac{a+b}{1+ab}, \tag{15.1}$$

where

$$a = \frac{\epsilon - \sqrt{\epsilon\mu}}{\epsilon + \sqrt{\epsilon\mu}} \quad \text{and} \quad b = e^{4i\pi\sqrt{\epsilon\mu}fd/c}. \tag{15.2}$$

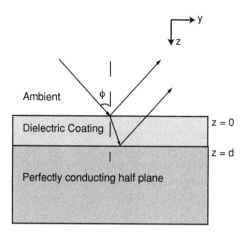

Figure 15.1: Interrogating high frequency wave impinging (angle of incidence ϕ) on coated (thickness d) perfectly conducting surface.

This expression can be derived directly from Maxwell's equation by considering the ratio of reflected to incident wave for example in the case of parallel polarized (TE_x) incident wave (see [BIKT, Jackson]).

An alternative and much more computationally intensive approach (which may be necessitated by some target geometries) employs the far field pattern for reflected waves computed directly using Maxwell's equations (see Example 4). In two dimensions, for a reflecting body Ω with coating layer Ω_1 and computational domain Π with an interrogating plane wave $E^{(i)}$, the scattered field $E^{(s)}$ satisfies the Helmholtz equation [CK1, CK2]

$$\nabla \cdot \left(\frac{1}{\mu}\nabla E^{(s)}\right) + \epsilon\omega^2 E^{(s)} = -\nabla \cdot \left(\frac{1}{\mu}\nabla E^{(i)}\right) - \epsilon\omega^2 E^{(i)} \qquad \text{in } \Pi \setminus \bar{\Omega}$$

$$E^{(s)} = -E^{(i)} \qquad \text{on } \partial\Omega$$

$$\left[\frac{1}{\mu}\frac{\partial E}{\partial n}\right] = [E] = 0 \qquad \text{on } \partial\Omega_1 \setminus \partial\Omega$$

$$\frac{\partial E^{(s)}}{\partial n} - ikE^{(s)} - \frac{i}{2k}\frac{\partial^2 E^{(s)}}{\partial s^2} = 0 \qquad \text{on } \partial\Pi$$

$$\frac{\partial E^{(s)}}{\partial s} - ik\frac{3}{2}E^{(s)} = 0 \qquad \text{at } C,$$

$$(15.3)$$

where the Silver-Müller radiation condition has been approximated by a second-order absorbing boundary condition on $\partial\Pi$ as described in [BJR, HKNT, HRT]. The vectors n and s denote the normal and tangential directions on the boundary $\partial\Pi$, respectively, and C is the set of the corner points of Π. Here $[\cdot]$ denotes the jumps at interfaces. The incident field in air is given by

$$E^{(i)} = e^{ik(x_1 \cos\alpha + x_2 \sin\alpha)},$$

where α is the *interrogation angle* $(\alpha = \frac{\pi}{2} - \phi)$ and $k = 2\pi f/c = \omega/c$ is the *interrogation wave number* corresponding to the interrogating frequency f. The corresponding far field pattern is given by [BIKT, CK1, CK2]

$$F_\alpha(\alpha + \pi; \epsilon, \mu, \alpha, f) =$$
$$\lim_{r \to \infty} \left(\sqrt{8\pi k r} \; e^{-i(kr + \pi/4)} \; E^{(s)}(r\cos(\alpha + \pi), r\sin(\alpha + \pi); \epsilon, \mu, \alpha, f) \right),$$
$$(15.4)$$

where $E^{(s)}(x_1, x_2; \epsilon, \mu, \alpha, f)$ is the scattered electromagnetic field. This can be used as a measure of the reflected field intensity instead of the reflection coefficient R of (15.1)–(15.2).

The evader and the interrogator are each subject to uncertainties as to the actions of the other. The evader wants to choose a best coating design (i.e., best ϵ's and μ's) while the interrogator wants to choose best angles of interrogation α and interrogating frequencies f. Each player must act in the presence of incomplete information about the other's action. Partial information regarding capabilities and tendencies of the adversary can be embodied in probability distributions for the choices to be made. That is, we may formalize this by assuming the evader may choose (with an as yet to be determined set of probabilities) dielectric permittivity and magnetic permeability parameters (ϵ, μ) from admissible sets $\mathcal{E} \times \mathcal{M}$ while the interrogator chooses angles of interrogation and interrogating frequencies (α, f) from sets $\mathcal{A} \times \mathcal{F}$. The formulation here is based on the *mixed strategies* proposals of von Neumann [Aubin, VN, VNM] and the ideas can be summarized as follows. The evader does not choose a single coating, but rather has a set of possibilities available for choice. He only chooses the probabilities with which he will employ the materials on a target. This, in effect, disguises his intentions from his adversary. By choosing his coatings randomly (according to a best strategy to be determined in, for example, a min-max game), he prevents adversaries from discovering which coating he will use—indeed, even he does not know which coating will be chosen for a given target. The interrogator, in a similar approach, determines best probabilities for choices of frequency and angle in the interrogating signals. Note that such a formulation tacitly assumes that the adversarial relationship persists with multiple attempts at evasion and detection.

The associated min-max problem consists of the evader choosing distributions $P_e(\epsilon, \mu)$ over $\mathcal{E} \times \mathcal{M}$ to minimize the reflected field (one can use either (15.1), in which case $\alpha = \frac{\pi}{2}$ is fixed, or (15.4); here we use (15.4) to illustrate ideas) while the interrogator chooses distributions $P_i(\alpha, f)$ over $\mathcal{A} \times \mathcal{F}$ to maximize the reflected field. If $B_R(\epsilon, \mu, \alpha, f)$ is the chosen measure of reflected field and $\mathcal{P}_e = \mathcal{P}(\mathcal{E} \times \mathcal{M})$ and $\mathcal{P}_i = \mathcal{P}(\mathcal{A} \times \mathcal{F})$ are the corresponding sets of probability distributions or measures over $\mathcal{E} \times \mathcal{M}$ and $\mathcal{A} \times \mathcal{F}$, respectively, then the cost functional involving (15.4) for the min-max problem can be defined by

$$J(P_e, P_i) = \int_{\mathcal{E} \times \mathcal{M}} \int_{\mathcal{A} \times \mathcal{F}} |B_R(\epsilon, \mu, \alpha, f)|^2 dP_e(\epsilon, \mu) dP_i(\alpha, f). \quad (15.5)$$

The problems thus formulated are special cases of classical *static zero–sum two player non-cooperative games* [Aubin, Basar] where the evader minimizes over $P_e \in \mathcal{P}_e$ and the interrogator maximizes over $P_i \in \mathcal{P}_i$. In such games one defines *upper* and *lower values* for the game by

$$\overline{J} = \inf_{P_e \in \mathcal{P}_e} \sup_{P_i \in \mathcal{P}_i} J(P_e, P_i)$$

and

$$\underline{J} = \sup_{P_i \in \mathcal{P}_i} \inf_{P_e \in \mathcal{P}_e} J(P_e, P_i).$$

The first represents a security level (worst case scenario) for the evader while the latter is a security level for the interrogator. It is readily argued that $\underline{J} \leq \overline{J}$ and if the equality $J^* = \underline{J} = \overline{J}$ holds, then J^* is called *the value* of the game. Moreover, if there exist $P_e^* \in \mathcal{P}_e$ and $P_i^* \in \mathcal{P}_i$ such that

$$J^* = J(P_e^*, P_i^*) = \min_{P_e \in \mathcal{P}_e} J(P_e, P_i^*) = \max_{P_i \in \mathcal{P}_i} J(P_e^*, P_i),$$

then (P_e^*, P_i^*) is a *saddle point solution* or *non-cooperative equilibrium* of the game.

To investigate theoretical, computational, and approximation issues for these problems, it is necessary to put a topology on the space of probability measures: a natural choice for both $\mathcal{P}_e$ and $\mathcal{P}_i$ is the Prohorov metric topology. Using its properties and arguments similar to those discussed earlier in Chapter 14 (see also [BBi, BG1, BG2, BK, BPi]), one can develop well-posedness and approximation results for the min-max problems defined above. Efficient computational methods that correspond to von Neumann's *finite mixed strategies* [Aubin] can readily be presented in this context. These can be based on several approximation theories that have been discussed above. In particular, either the DEL(M) or SPL(M,N) methods discussed in the context of Theorems 14.3 and 14.4 of Chapter 14 can be used.

15.1.2 Theoretical Results

To establish existence of a saddle point solution for the evasion-interrogation problems formulated above, one can employ a fundamental result of von Neumann [VN, VNM] as stated by Aubin (see [Aubin, p. 126]).

Theorem 15.1 (von Neumann) *Suppose X_0, Y_0 are compact, convex subsets of metric linear spaces X, Y, respectively. Further suppose that*

(i) *for all $y \in Y_0$, $x \to f(x, y)$ is convex and lower semi-continuous;*

(ii) *for all $x \in X_0$, $y \to f(x, y)$ is concave and upper semi-continuous.*

Then there exists a saddle point (x^*, y^*) such that

$$f(x^*, y^*) = \min_{X_0} \max_{Y_0} f(x, y) = \max_{Y_0} \min_{X_0} f(x, y).$$

A straightforward application of these results to the problems under discussion (where J of (15.5) is linear in P_e and P_i) leads immediately to desired well-posedness results.

Theorem 15.2 *Suppose $\mathcal{E}, \mathcal{M}, \mathcal{A}, \mathcal{F}$ are compact and the spaces $X_0 = \mathcal{P}(\mathcal{E} \times \mathcal{M})$, $Y_0 = \mathcal{P}(\mathcal{A} \times \mathcal{F})$ are taken with the Prohorov metric. Then X_0, Y_0 are compact, convex subsets of $X = C_B^*(\mathcal{E} \times \mathcal{M})$ and $Y = C_B^*(\mathcal{A} \times \mathcal{F})$, respectively. Moreover, there exists $(P_e^*, P_i^*) \in \mathcal{P}(\mathcal{E} \times \mathcal{M}) \times \mathcal{P}(\mathcal{A} \times \mathcal{F})$ such that*

$$J(P_e^*, P_i^*) = \min_{\mathcal{P}(\mathcal{E} \times \mathcal{M})} \max_{\mathcal{P}(\mathcal{A} \times \mathcal{F})} J(P_e, P_i) = \max_{\mathcal{P}(\mathcal{A} \times \mathcal{F})} \min_{\mathcal{P}(\mathcal{E} \times \mathcal{M})} J(P_e, P_i).$$

For computations, we may use the "delta" approximations DEL(M) based on Theorem 14.3 and Lemma 14.1 and obtain

$$P_e(\epsilon, \mu) \approx \sum p_e^j \Delta_{(\epsilon_j, \mu_j)}(\epsilon, \mu),$$
$$P_i(\alpha, f) \approx \sum p_i^j \Delta_{(\alpha_j, f_j)}(\alpha, f),$$

or

$$dP_e(\epsilon, \mu) \approx \sum p_e^j \delta_{(\epsilon_j, \mu_j)}(\epsilon, \mu) d\epsilon d\mu,$$
$$dP_i(\alpha, f) \approx \sum p_i^j \delta_{(\alpha_j, f_j)}(\alpha, f) d\alpha df.$$

As noted above, a convergence theory can be found in [BBi]. This formulation corresponds precisely to von Neumann's finite mixed strategies framework for protection by disguising intentions from opponents, i.e., by introducing uncertainty in the players' choices.

To illustrate the computational framework based on the delta measure approximations, we take

$$dP_e^M(\epsilon, \mu) = \sum_{j=1}^M p_j^M \delta_{(\epsilon_j^M, \mu_j^M)} d\epsilon d\mu \quad \text{and} \quad dP_i^N(\alpha, f) = \sum_{k=1}^N q_k^N \delta_{(\alpha_k^N, f_k^N)} d\alpha df,$$

which can be represented respectively by

$$\bar{p}^M = \{p_j^M\}_{j=1}^M \in P^M \equiv \{\bar{p} \in \mathbb{R}^M \mid p_j \geq 0, \sum_{j=1}^M p_j = 1\} \qquad (15.6)$$

and

$$\bar{q}^N = \{q_k^N\}_{k=1}^N \in Q^N \equiv \{\bar{q} \in \mathbb{R}^N \mid q_k \geq 0, \sum_{k=1}^N q_k = 1\}. \qquad (15.7)$$

We note that in this case the p_j^M, q_k^N are the probabilities associated with the use of the material parameters (ϵ_j^M, μ_j^M) and interrogating parameters (α_k^N, f_k^N) in the mixed strategies of the evader and interrogator, respectively.

Then $J(P_e^M, P_i^N)$ reduces to

$$\mathcal{J}(\bar{p}^M, \bar{q}^N) = \sum_{k=1}^{N} \sum_{j=1}^{M} p_j^M \left| B_R(\epsilon_j^M, \mu_j^M, \alpha_k^N, f_k^N) \right|^2 q_k^N$$

where B_R is a measure of the reflected field (either the reflection coefficient or the far field scattering intensity). Since P^M, Q^N are compact, convex subsets of $\mathbb{R}^M, \mathbb{R}^N$, respectively, we have the following theorem.

Theorem 15.3 *For fixed M, N there exists $(\bar{p}_*^M, \bar{q}_*^N)$ in $P^M \times Q^N$ such that*

$$
\begin{aligned}
\mathcal{J}^* &= \mathcal{J}(\bar{p}_*^M, \bar{q}_*^N) = \min_{\bar{p}^M \in P^M} \max_{\bar{q}^N \in Q^N} \mathcal{J}(\bar{p}^M, \bar{q}^N) \\
&= \max_{\bar{q}^N \in Q^N} \min_{\bar{p}^M \in P^M} \mathcal{J}(\bar{p}^M, \bar{q}^N).
\end{aligned}
$$

Assume further that $(\epsilon, \mu, \alpha, f) \to B_R(\epsilon, \mu, \alpha, f)$ is continuous on $\mathcal{E} \times \mathcal{M} \times \mathcal{A} \times \mathcal{F}$ which is assumed compact. Then there exists a sequence $(\bar{p}_*^M, \bar{q}_*^N)$ of elements from $P^M \times Q^N$, respectively, with corresponding (P_e^M, P_i^N) in $\mathcal{P}(\mathcal{E} \times \mathcal{M}) \times \mathcal{P}(\mathcal{A} \times \mathcal{F})$ converging in the Prohorov metric to (P_e^*, P_i^*) which is a saddle point for the original min-max problem.

We give the arguments for these results since they are in some sense representative of the use of functional analysis in such problems. The hypothesis that the map $(\epsilon, \mu, \alpha, f) \to B_R(\epsilon, \mu, \alpha, f)$ be continuous on $\mathcal{E} \times \mathcal{M} \times \mathcal{A} \times \mathcal{F}$ implies continuity of the map $(P_e, P_i) \to J(P_e, P_i)$ on $X_0 \times Y_0 \to \mathbb{R}^1_+$, where $X_0 = \mathcal{P}(\mathcal{E} \times \mathcal{M})$ and $Y_0 = \mathcal{P}(\mathcal{A} \times \mathcal{F})$ are compact convex subsets of $X = C_B^*(\mathcal{E} \times \mathcal{M})$ and $Y = C_B^*(\mathcal{A} \times \mathcal{F})$, respectively. This is due to the compactness of $\mathcal{E} \times \mathcal{M}$ and $\mathcal{A} \times \mathcal{F}$, and properties of the Prohorov metric.

Let P^M, Q^N be defined as in (15.6),(15.7), and observe that these are compact convex subsets of $\mathbb{R}^M, \mathbb{R}^N$, respectively. Now let $P_e \in X_0$ and $P_i \in Y_0$ be arbitrary. Then by Theorem 14.3 and (iii) of Lemma 14.1 above (see also [BBi]), there exists a sequence $\bar{p}^M, \bar{q}^N \in P^M, Q^N$ with associated measures P_e^M, P_i^N, respectively, such that

$$P_e^M \to P_e \text{ in } X_0 \text{ as } M \to \infty \text{ and } P_i^N \to P_i \text{ in } Y_0 \text{ as } N \to \infty.$$

Let P_e^{M*}, P_i^{N*} (guaranteed to exist by continuity, compactness, and the von Neumann theorem) with coordinates $\bar{p}_*^M, \bar{q}_*^N$, respectively, satisfy

$$J(P_e^{M*}, P_i^N) \leq J(P_e^{M*}, P_i^{N*}) \leq J(P_e^M, P_i^{N*}) \qquad (15.8)$$

for all (P_e^M, P_i^N) in $X_0 \times Y_0$ with coordinate representations in P^M, Q^N, respectively. Now $X_0 \times Y_0$ compact implies that there exists a subsequence $(P_e^{M_k*}, P_i^{N_k*})$ and $(\tilde{P}_e, \tilde{P}_i)$ in $X_0 \times Y_0$ such that $(P_e^{M_k*}, P_i^{N_k*}) \to (\tilde{P}_e, \tilde{P}_i)$ in $X_0 \times Y_0$.

Consider (15.8) for indices M_k, N_k, so that

$$J(P_e^{M_k*}, P_i^{N_k}) \le J(P_e^{M_k*}, P_i^{N_k*}) \le J(P_e^{M_k}, P_i^{N_k*}).$$

Then given the continuity of J and taking the limit we find

$$J(\tilde{P}_e, P_i) \le J(\tilde{P}_e, \tilde{P}_i) \le J(P_e, \tilde{P}_i).$$

Since P_e, P_i are arbitrary in X_0, Y_0, respectively, we have $(\tilde{P}_e, \tilde{P}_i)$ is a saddle point for the original min-max problem.

Further note that since these arguments hold for any subsequence of (P_e^{M*}, P_i^{N*}), then

$$J(P_e^{M*}, P_i^N) \le J(P_e^{M*}, P_i^{N*}) \le J(P_e^M, P_i^{N*})$$

and $P_e^{M*} \to \tilde{P}_e$ as well as $P_i^{N*} \to \tilde{P}_i$. That is, any subsequence of (P_e^{M*}, P_i^{N*}) has a convergent subsequence to $(\tilde{P}_e, \tilde{P}_i)$ so that the sequence itself must converge to $(\tilde{P}_e, \tilde{P}_i)$.

15.2   Optimal Design Techniques

If the parameters (diffusion, transport, growth/birth/death rates, dielectric permittivity and magnetic permeability, Young's modulus, damping coefficients, etc.)   in the mathematical models discussed throughout this presentation are known, the models can be used for simulation, prediction, control design, etc. However, typically one does not have accurate values for the parameters. Instead, one must estimate the parameters using experimental data. This leads to parameter estimation or inverse problems that have already been discussed in some detail in this book. A major question that both experimentalists and inverse problem investigators often face is how to best collect the data to enable one to efficiently and accurately estimate model parameters. This is the well-known and widely studied *optimal design* problem.

Traditional optimal design methods (D-optimal, E-optimal, c-optimal) [AD, BW, Fed, FedHac] use information from the model to find the sampling distribution or mesh for the observation times (and/or locations in spatially distributed problems) that minimizes a design criterion, quite often a function of the Fisher Information Matrix (FIM). Experimental data taken on this optimal mesh is then expected to result in more accurate parameter estimates.

One may formulate the optimal design problem in the context of general optimization problems over distributions of sampling times. A number of optimal design techniques are available; here we illustrate how three of these can be readily formulated in terms of a Prohorov metric optimization framework that guarantees well-posedness and leads to viable computational approaches. The optimal design methods examined are SE-optimal, D-optimal, and E-optimal design. SE-optimal design (standard error optimal design) was introduced in [BDEK] and subsequently investigated in [BHK]. The goal of SE-optimal design is to find the observation times $\tau = \{t_i\}$ that minimize the sum of squared normalized standard errors of the estimated parameters as defined by asymptotic distribution results from statistical theories [BDSS, BHR, DG, SeWi]. D-optimal and E-optimal design methods minimize functions of the covariance in the parameter estimates [AD, BW, FedHac]. In D-optimal design one seeks to find the mesh that minimizes the volume of the confidence interval ellipsoid of the asymptotic covariance matrix, while in E-optimal design the goal is to minimize the largest principal axis of the confidence interval ellipsoid of the asymptotic covariance matrix. One approach to comparing these design criteria [BHK] involves comparison of the resulting standard errors for the estimated parameters. In this case one expects that SE-optimal design will result in smaller standard errors compared with the other optimal design methods since SE-optimal design optimizes directly on the standard errors themselves while the D-optimal and E-optimal methods minimize other functions related to the standard errors. This is true to some extent but other considerations as discussed in [BHK] are pertinent.

15.2.1   Optimal Design Formulations

Following [BDEK, BHK], we may introduce a formulation of *ideal* inverse problems in which continuous in time observations are available; while not practical, the associated considerations provide valuable insight. A major question in this context is how to choose sampling distributions in an intelligent manner. Indeed, this is the fundamental question treated in the optimal design literature and methodology.

Underlying our considerations is a *mathematical model* (which here we describe in the context of general nonlinear differential equations although other dynamical systems, e.g., partial or delay differential equations, could easily be employed)

$$
\begin{aligned}
\dot{w}(t) &= g(t, w(t), q), \\
w(0) &= w_0, \\
f(t, \theta) &= C(w(t, \theta)), \quad t \in [0, T],
\end{aligned}
\tag{15.9}
$$

where $w(t) \in \mathbb{R}^n$ is the vector of state variables of the system, $f(t, \theta) \in \mathbb{R}^m$ is the vector of observable or measurable outputs, $q \in \mathbb{R}^r$ are the system parameters, $\theta = (q, w_0) \in \mathbb{R}^p, p = r + n$ is the vector of system parameters plus initial conditions w_0, while g and C are mappings $\mathbb{R}^{1+n+r} \to \mathbb{R}^n$ and $\mathbb{R}^n \to \mathbb{R}^m$, respectively. To consider measures of uncertainty in estimated parameters, one also requires a *statistical model* [BDSS]. Our statistical model is given by the stochastic process

$$
Y(t) = f(t, \theta_0) + \mathcal{E}(t).
\tag{15.10}
$$

Here $\mathcal{E}$ is a noisy random process representing measurement errors and, as usual in statistical formulations [BDSS, BDEK, DG, SeWi], θ_0 is a hypothesized "true" value of the unknown parameters. We make the following standard assumptions on the random variable $\mathcal{E}(t)$:

$$
\begin{aligned}
E(\mathcal{E}(t)) &= 0, \quad t \in [0, T], \\
\mathrm{Var}\mathcal{E}(t) &= \sigma(t)^2 I, \quad t \in [0, T], \\
\mathrm{Cov}(\mathcal{E}(t)\mathcal{E}(s)) &= \sigma(t)^2 I \delta(t - s), \quad t, s \in [0, T],
\end{aligned}
$$

where $\delta(s) = 1$ for $s = 0$ and $\delta(s) = 0$ for $s \neq 0$. A realization of the observation process is given by

$$
y(t) = f(t, \theta_0) + \varepsilon(t), \quad t \in [0, T],
$$

where the measurement error $\varepsilon(t)$ is a realization of $\mathcal{E}(t)$.

We introduce a generalized weighted least squares criterion

$$
J(y, \theta) = \int_0^T \frac{1}{\sigma(t)^2} |y(t) - f(t, \theta)|^2 \, dP(t),
\tag{15.11}
$$

where P is a general measure on $[0, T]$. We seek the parameter estimate $\hat{\theta}$ by minimizing $J(y, \theta)$ for θ. Since P represents a weighting of the difference between data and model output, we may assume, as in the treatment of non-cooperative min-max games above, that P is a bounded measure on $[0, T]$.

If, for points $\tau = \{t_i\}$, $t_1 < \cdots < t_M$ in $[0, T]$, we take

$$P_\tau = \sum_{i=1}^{M} \Delta_{t_i}, \tag{15.12}$$

where we recall that Δ_a denotes the Dirac delta distribution with atom at $\{a\}$, we obtain

$$J_d(y, \theta) = \sum_{i=1}^{M} \frac{1}{\sigma(t_i)^2} |y(t_i) - f(t_i, \theta)|^2, \tag{15.13}$$

which is the weighted least squares cost functional for the case where we take a finite number of measurements in $[0, T]$ (note that if we want to obtain probability measures, we should use $\frac{1}{M} P_\tau$ in the formulations). Of course, the introduction of the measure P allows us to change the weights in (15.13) or the weighting function in (15.11). For instance, if P is absolutely continuous with density $p(\cdot)$ the error functional (15.11) is just the weighted L_2-norm of $y(\cdot) - f(\cdot, \theta)$ with weight $p(\cdot)/\sigma(\cdot)^2$.

To facilitate our discussions we introduce the *Generalized Fisher Information Matrix* (GFIM)

$$F(P, \theta_0) \equiv \int_0^T \frac{1}{\sigma^2(s)} \nabla_\theta^\mathsf{T} f(s, \theta_0) \nabla_\theta f(s, \theta_0) \, dP(s), \tag{15.14}$$

where ∇_θ is a row vector given by $(\partial_{\theta_1}, \ldots, \partial_{\theta_p})$ and hence $\nabla_\theta f$ is an $m \times p$ matrix. It follows that the usual discrete FIM corresponding to P_τ as in (15.12) is given by

$$F(\tau) = F(P_\tau, \theta_0) = \sum_{j=1}^{M} \frac{1}{\sigma^2(t_j)} \nabla_\theta f(t_j, \theta_0)^\mathsf{T} \nabla_\theta f(t_j, \theta_0). \tag{15.15}$$

Subsequently we simplify notation and use $\tau = \{t_i\}$ to represent the dependence of $P = P_\tau$ on τ when it has the form (15.12). When one chooses P as simple Lebesgue measure then the GFIM reduces to the continuous FIM

$$F_C = \int_0^T \frac{1}{\sigma^2(s)} \nabla_\theta f(s, \theta_0)^\mathsf{T} \nabla_\theta f(s, \theta_0) \, ds. \qquad (15.16)$$

The major question in optimal design of experiments is how to best choose P in some family $\mathcal{P}(0, T)$ of observation distributions.

The introduction of the measure P above allows for a unified framework for optimal design criteria which incorporates the three design criteria mentioned above. As already noted, the GFIM $F(P, \theta)$ introduced in (15.14) depends critically on the measure P. We also remark that we can, without loss of generality, further restrict ourselves to probability measures on $[0, T]$. Thus, let $\mathcal{P}(0, T)$ denote the set of all probability measures on $[0, T]$ and assume that a functional $\mathcal{J} : \mathbb{R}^{p \times p} \to \mathbb{R}^+$ of the GFIM is given. The *optimal design problem* associated with $\mathcal{J}$ is one of finding a probability measure $\hat{P} \in \mathcal{P}(0, T)$ such that

$$\mathcal{J}\big(F(\hat{P}, \theta_0)\big) = \min_{P \in \mathcal{P}(0,T)} \mathcal{J}\big(F(P, \theta_0)\big). \qquad (15.17)$$

A general theoretical framework for existence and approximation in the context of $\mathcal{P}(0, T)$ taken with the Prohorov metric is given for these problems in [BDEK, BHK] using the ideas discussed in Chapter 14. In particular, this theory permits development of computational methods using weighted discrete measures (i.e., weighted versions of (15.12)).

15.2.2 Theoretical Summary

To summarize and further develop the theoretical considerations that are the basis of our considerations here, we first recall from our presentation above that $(\mathcal{P}(0, T), \rho)$ is a complete, compact, and separable metric space. (We will in this section just denote this space by $\mathcal{P}(0, T)$ since the ρ will be understood.)

A first observation is that the GFIM as defined in (15.14) is ρ continuous on $\mathcal{P}(0, T)$ for problems in which the observation functions $f(\cdot, \theta)$ are continuously differentiable on $[0, T]$. Thus, whenever $\mathcal{J} : \mathbb{R}^{p \times p} \to \mathbb{R}^+$ is continuous we have that $P \to \mathcal{J}(F(P, \theta))$ is continuous from $\mathcal{P}(0, T)$ to $\mathbb{R}^+$. Since $\mathcal{P}(0, T)$ is ρ compact, we obtain immediately the existence of solutions for the optimization problems

$$\hat{P}_{\mathcal{J}} \equiv \arg \min_{P \in \mathcal{P}(0,T)} \mathcal{J}(F(P, \theta_0)). \qquad (15.18)$$

A second observation follows from Theorem 14.3 above; we found in particular the density in $\mathcal{P}(0, T)$ of finite convex combinations over rational coefficients of Dirac measures Δ_a with atoms at a. Specifically, we have for $\mathcal{T}_0 = \{t_j\}_{j=1}^{\infty}$ a given countable, dense subset of $[0, T]$, that the set

$$\mathcal{P}_0(0,T) := \left\{ P \in \mathcal{P}(0,T) \middle| P = \sum_{j=1}^{k} p_j \Delta_{t_j}, \; k \in \mathbb{N}^+, \; t_j \in \mathcal{T}_0, \; p_j \geq 0, \right.$$

$$\left. p_j \text{ rational}, \; \sum_{j=1}^{k} p_j = 1 \right\}$$

is dense in $\mathcal{P}(0,T)$ in the Prohorov metric ρ. In short, the set of $P \in \mathcal{P}(0,T)$ with finite support in $\mathcal{T}_0$ and rational masses is dense in $\mathcal{P}(0,T)$. This leads, for a given choice $\mathcal{J}$, to approximation schemes for $\hat{P}_{\mathcal{J}}$ as defined in (15.18). To implement these for a given choice of $\mathcal{J}$ (examples are discussed below) would require approximation by $P^M_{\{p_j,t_j\}} = \sum_{j=1}^{M} p_j \Delta_{t_j}$ in the GFIM (15.14) and then optimization over appropriate sets of $\{p_j, t_j\}$ in (15.18) with P replaced by $P^M_{\{p_j,t_j\}}$. For a fixed M, existence of minima in these problems follows from the theory outlined previously. In standard optimal designs these problems are approximated even further by fixing the weights or masses p_j as $p_j = \frac{T}{M}$ (which then becomes simply a scale factor in the sum) and searching over the $\{t_j\}$. This, of course, is equivalent to replacing the $P^M_{\{p_j,t_j\}}$ by P_τ of (15.12) in (15.14) and searching over the $\tau = \{t_j\}$ for a fixed number M of grid points. This embodies the tacit assumption of equal value of the observations at each of the times $\{t_j\}$. We observe that weighting of information at each of the observation times is carried out in the inverse problems via the weights $\sigma(t_j)$ for observation variances in (15.13). We further observe that the weights $\{p_j\}$ in $P^M_{\{p_j,t_j\}}$ are related to the value of the observations as a function of the model sensitivities $\nabla_\theta f(t_j, \theta_0)$ in the FIM while the weights $\frac{1}{\sigma(t_j)^2}$ are related to the reliability in the data measurement processes. We note that all of our remarks on theory related to existence above in the general probability measure case also hold for this discrete minimization case.

15.2.3   Design Strategy Examples

As we have already noted, the formulation (15.18) incorporates all strategies for optimal design which try to optimize a functional depending continuously on the elements of the Fisher information matrix. In case of the traditional design criteria mentioned in the introduction, $\mathcal{J}$

is the determinant (D-optimal), the smallest eigenvalue (E-optimal), or a quadratic form (c-optimal) of the inverse of the Fisher information matrix. Specifically, this includes the optimal design methods we discuss here: SE-optimal design, D-optimal design, and E-optimal design. The design cost functional for the SE-optimal design method is given by (see [BDEK, BHK])

$$\mathcal{J}_{SE}(F) = \sum_{i=1}^{p} \frac{1}{\theta_{0,i}^2}(F^{-1})_{ii}, \qquad (15.19)$$

where $F = F(\tau)$ is the FIM, defined above in (15.15), θ_0 is the true parameter vector, and p is the number of parameters to be estimated. Note that both inversion and taking the trace of a matrix are continuous operations. We observe that $F_{ii}^{-1} = SE_i(\theta_0)^2$. Therefore, SE-optimal design minimizes the sum of squared normalized standard errors.

D-optimal design minimizes the volume of the confidence interval ellipsoid for the covariance matrix ($\Sigma_0^M = F^{-1}$). The design cost functional for D-optimal design is given by (see [BW, FedHac])

$$\mathcal{J}_D(F) = \det(F^{-1}). \qquad (15.20)$$

Again we note that taking the determinant is a continuous operation on matrices so that $\mathcal{J}_D$ is continuous in F as required by the theory.

E-optimal design minimizes the principal axis of the confidence interval ellipsoid of the covariance matrix (defined in the asymptotic theory summarized in [BDSS]). The design cost functional for E-optimal design is given by (see [AD, BW])

$$\mathcal{J}_E(F) = \max_i \frac{1}{\lambda_i}, \qquad (15.21)$$

where λ_i, $i = 1 \ldots p$ are the eigenvalues of F (which are continuous functions of F). Therefore $\frac{1}{\lambda_i}$, $i = 1 \ldots p$, corresponds to the eigenvalues of the asymptotic covariance matrix $\Sigma_0^M = F^{-1}$.

15.3 Generalized Curves and Relaxed Controls of Variational Theory

The weak* topology (as characterized by the Prohorov metric) on measures when they are embedded in the topological dual of the space of continuous functions $(\mathcal{P}(Q) \subset C_B^*(Q))$ has played a fundamental role in the calculus of variations dating to the early contributions on generalized curves by L.C. Young [Young37, Young38] and E.J. McShane [McShane40-1, McShane40-2, McShane67]. With the emergence of control theory (roughly a reformulation of the calculus of variations wherein the constraints are explicit dynamical systems) in the 1960s, similar topologies were used in the study of "sliding regimes" by A.F. Filippov [Filippov62] and "relaxed controls" by J. Warga [Warga62, Warga67, Warga72]. The underlying ideas in both settings resulted from a lack of closure of trajectories in ordinary function spaces, e.g., think of Sobolev spaces and distributions (Schwartz) employed in treating "weak" solutions of partial differential equations. To circumvent the difficulties due to lack of closure, one extends the concept of "function" to allow for curves or solutions in a more general sense. To be a little more precise, consider the curves depicted in Figure 15.2 where one has a curve whose derivative is piecewise constant approximating the smoother curve over a fixed control interval $[0, T]$. In the case of control input to a dynamical system, the first curve corresponds to a differential equation with piecewise changing constant inputs. As one increases the frequency of change in the piecewise constant input or control, one has a very high frequency in control switches, hence the introduction of the concept of "chattering controls" in the control field.

Figure 15.2: Relaxed or generalized curves.

This can be succinctly described in terms of the dynamical system (the constraints) in control problems. Consider a control system

$$\dot{w}(t) = g(t, w(t), u(t)) \qquad (15.22)$$

with a control input $t \to u(t)$ having values in a given control constraint set U with piecewise constant controls switching very often over a period of time. The corresponding trajectory can be represented by a weighted convex combination of trajectories satisfying

$$\dot{w}(t) = \sum p_j g(t, w(t), u_j). \tag{15.23}$$

In the limit as the number of switches increases, the derivatives approach generalized limits of convex combinations of delta functions that can be represented by

$$\dot{w}(t) = \lim \sum p_j g(t, w(t), u_j). \tag{15.24}$$

This becomes, in light of the approximation of Theorem 14.3,

$$\dot{w}(t) = \lim \sum p_j g(t, w(t), u_j) \approx \int_U g(t, w(t), u) d\mu(t, u) \tag{15.25}$$

$$= \mathcal{M}[g(t, w(t), \cdot); t] \tag{15.26}$$

where $\mathcal{M}[\cdot; t]$ is an *averaging operator* over U. The function $\mathcal{M}[\cdot; t]$ is essentially a time dependent probability measure which can be identified with elements in C^* by

$$\mathcal{M}[\cdot; t] \iff \mu(t, \cdot) \text{ in } C^*(U),$$

through

$$\mathcal{M}[\phi(\cdot); t] = \int_U \phi(u) d\mu(t, u)$$

for $\phi \in C(U)$, $\mu(t, \cdot) \in \mathcal{P}(U) \subset C^*(U)$, where $\mathcal{P}(U)$ is the set of probability measures on U.

We recall that Prohorov convergence is equivalent to weak* convergence on $\mathcal{P}(U) \subset C_B^*(U)$. As before, we denote the metric space $(\mathcal{P}(U), \rho)$ taken with the Porhorov metric simply by $\mathcal{P}(U)$. The corresponding generalized control problem becomes:

Minimize

$$J(w(\mu), \mu)$$

over

$$\mu(t, \cdot) \in \mathcal{P}(U) \subset C_B^*(U) \quad \text{or} \quad \mu \in L_\infty(0, T; \mathcal{P}(U))$$

subject to

$$\dot{w}(t) = \int_U g(t, w(t), u) d\mu(t, u)$$

$$= \mathcal{M}[g(t, w(t), \cdot); t] \tag{15.27}$$

Thus, controls μ or $\mathcal{M}$ are taken in $L_\infty(0,T;\mathcal{P}(U)) \subset L_1(0,T;C_B(U))^*$ $= L_\infty(0,T;C_B^*(U))$ where, as noted above, one uses the Prohorov or weak* topology on $\mathcal{P}(U)$ (recall from Section 4.3 that $L_\infty(0,T;X^*) = L_1(0,T;X)^*$); see [Warga72, Warga67, ChapterIV] for details.

The above formulation convexifies and generalizes the original problem. We obtain closure, and hence existence through the compactness inherited from the Prohorov metric — see Chapter 14. In this formulation the corresponding trajectories are called relaxed trajectories or generalized trajectories (in which the family of ordinary trajectories are embedded densely — again recall Theorem 14.3). This results in existence (inf = min) of optimal controls in a wider class of functions: the relaxed controls $\mu(t,\cdot) \in \mathcal{P}(U)$. McShane showed that under not too restrictive assumptions, an infimum for the problem is actually attained by a generalized control that happens to be ordinary, i.e., the optimal measure is actually equivalent to a discrete (Dirac) measure at a single atom $u(t) \in U$ at each t. With a slightly different outcome, Warga showed that knowledge of a minimizing generalized curve permits approximation with nearby ordinary curves (again the density in Theorem 14.3 suggests this!). Of course, these contributors did not know of the Prohorov metric compactness results or the density results of Theorem 14.3—we note that the Prohorov metric was introduced in 1956 (*after* the early results of Young, McShane) and moreover in another field of mathematical sciences.

15.4   Preisach Hysteresis in Smart Materials

Another area in which the weak* or Prohorov metric topology on measures plays an important role is in control of structures using hysteretic smart materials such as shape memory alloys, piezoelectrics, and magnetostrictives [BSW]. The input or control operators are often described in terms of Preisach hysteretic inputs. Modeling of Preisach hysteresis [Preisach35] also plays an important role in modeling of viscoelastic materials such as rubber, polymers, living tissues, as well as in modeling polarization and conductivity in electromagnetics.

As we shall see below, equations for controlled structures can be written in abstract operator form as

$$\ddot{y}(t) + D(t)\dot{y}(t) + A(t)y(t) = B(u)(t), \quad \text{in } V^*, \quad t \in (0,T); \quad (15.28)$$

where $B(u)$ is the control input term resulting from, in this case, a smart material hysteretic actuator. In [BanksKurdila, BKurW1,

BKurW2], the control term has the form

$$B(u, f)(t) = B_\mu(u, f)(t) \cong \int_{\bar{S}} \kappa_{\bar{s}}(u, f(\bar{s}))(t) d\mu(\bar{s}) \qquad (15.29)$$

where $B_\mu(u, f) \in L_2(0, T; V^*)$ involves a generalized smoothed Preisach-Krasnoselskii-Pokrovskii [BroSpre, KP, Ma, Vis] hysteretic control input kernel $\kappa_{\bar{s}}$. The properties of compactness in the weak* or Prohorov topology of the underlying input term play a fundamental role in well-posedness of the parameter estimation or inverse problems treated in [BanksKurdila, BKurW1, BKurW2]. We briefly outline the fundamentals of Preisach formulations as given in [BanksKurdila, BKurW1, BKurW2], where the authors used these hysteretic control influence operators representing smart material actuators to treat identification and approximation.

The hysteresis is embodied in the stress-strain laws $\sigma = \sigma(\varepsilon)$. The basic Preisach construction [Ma] is the simple ideal relay $k_{\bar{s}}(\varepsilon, \xi)$ as depicted in Figure 15.3. Here ξ is a switching variable that takes on values in $\{-1, 1\}$. The fundamental idea due to Preisach is to use multiple superimposed ideal relays in parallel connection as depicted in Figure 15.4. In terms of the stress-strain law one would then write

$$[\sigma(\varepsilon)](t) = \sum_j \alpha_j k_{\bar{s}_j}(\varepsilon, \xi)(t)$$

where $\bar{s}_j = (s_{j1}, s_{j2})$.

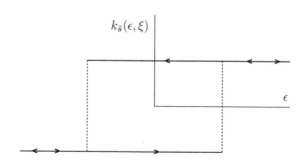

Figure 15.3: Simple ideal relay with $\xi \in \{-1, 1\}$.

While these superimposed relays as well as related Krasnoselskii-Pokrovskii kernels [KP] are useful conceptually, they lack a smoothness

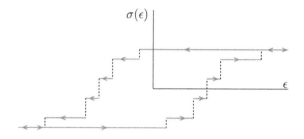

Figure 15.4: Superimposed multiple relays.

needed in many applications. In particular, the efforts on inverse problems involving estimation of measures in control input operators for well-posedness and approximation, as well as computational methods pursued in [BanksKurdila, BKurW1, BKurW2] required a generalization of the Preisach-Krasnoselskii-Pokrovskii theories, operators, and kernels. These authors combined smoothed relays for a continuum of switch points $\bar{s} = (s_1, s_2)$ to obtain the smoothed operators as depicted in Figure 15.5. The corresponding stress-strain input law can then be written

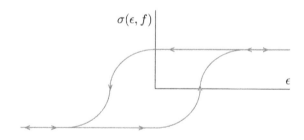

Figure 15.5: Smoothed relay.

$$\sigma(\varepsilon, f)(t) = \int_{\bar{S}_\Delta} \kappa_{\bar{s}}(\varepsilon, f(\bar{s}))(t) d\mu(\bar{s})$$

where $\bar{S}_\Delta$ is the Preisach domain depicted in Figure 15.6, $\mu \in \mathcal{P}(\bar{S}_\Delta)$ and f is a Borel measurable function taking on values in $\{-1, 1\}$. See [BKurW1, BKurW2] for details where the Prohorov metric in the form of weak* convergence plays a fundamental role.

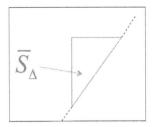

Figure 15.6: Preisach domain.

15.5 NPML and Mixing Distributions in Statistical Estimation

The overriding problem in Chapter 14 is the estimation of a probability distribution P from data which does not involve sampling directly from the distribution itself. These problems and the underlying mathematical constructs are very much related to fundamental topics in statistics and probability theory involving random effects, mixing distributions, and Nonparametric Maximum Likelihood (NPML) estimation methods [DG, DavidianGallant, DavidianGallant2, GallantNychka, Mallet, Mallet2, Lindsay, Lindsay2, Schumitzky] where some of the original motivating examples involved Physiologically-Based Pharmacokinetic (PBPK) problems described by ordinary differential equation models for individual patients. These problems are somewhat different in nature than those motivating the PMF ideas described above. The PMF techniques were developed for problems in which only aggregate data are available to use with either individual dynamics or aggregate dynamics while the primary formulation in hierarchical modeling and mixing distributions involves individual dynamics and individual longitudinal data. In particular, one usually assumes in the simplest formulation that one has N patients with m observations each (contrast this with the PMF formulation of Section 14.1 where one has a total number, say N, of observations which cannot be identified with particular individuals. The statistical literature has functional analytic foundations in both the efforts of Lindsay based on convex geometry [Lindsay, Lindsay2] and the efforts of Mallet and colleagues [Mallet, Mallet2] which involve convex analysis along with optimal design ideas as outlined above. In particular in [Mallet], the author develops the connection between maximizing the likelihood function and maximizing the determinant of the Fisher Information Matrix. Thus, Mallet connects the desired maximum likelihood problem to a D-optimal design problem. In fact, by invoking the design literature, Mallet is able to show that for a given data set taken from N individuals, an optimal discrete probability distribution (as developed in Section 14.1 above) exists that is composed of at most N atoms and weights, i.e., a combination of one-point designs (Dirac distributions). This involves a classic theorem of Carathéodory's on representation of boundary points of closed convex sets. This is the Nonparametric MLE (NPML) which, for N sampled individuals, can be realized as a discrete probability distribution containing $M \leq N$ points. Mallet also provides a computational algorithm for approximating this optimal design. We remark that, as in the PMF, the NPML is capable of the simultaneous estimation of a statistical structural parameter as well as a distribution over the dynamic model parameters.

In general it is difficult to compare the PMF and NPML formulations directly even if one assumes $m = 1$ (sampling without replacement) in a population; nonetheless, one can in some examples compare the NPML formulations and results of Mallet and those of the combined PMF and optimal designs of Sections 14.1 and 15.2—see [BKT1, BKT2]. We remark that, in the MLE literature, the parameter q is generally assumed to be in a subset of Euclidean space whereas in Example 5 of Chapter 14 illustrating the PMF techniques one has the measures taken over functional families of growth rates $\mathcal{G}$.

16 Control Theory for Distributed Parameter Systems

16.1 Motivation

Many of the functional analysis ideas and results presented in this book provide the essential foundations for development of modern *control theory* for so-called *Distributed Parameter Systems* (DPS). Control of DPS is the engineering terminology used primarily for control of partial differential equations and delay differential equations, examples of which we have discussed in previous chapters of this book. The research literature contains a plethora of control theory presentations in a Hilbert space framework which rely in a fundamental way on some of the topics we have discussed including strongly continuous and analytic semigroups, unbounded operator inputs, weak formulations via sesquilinear forms, adjoint operators, and finite element approximations via the Trotter-Kato theorems in Hilbert and Banach spaces. Our formulations in previous chapters of partial differential and delay differential equations as abstract infinite dimensional differential equations in either weak form or with semigroup representations are precisely the setting for much of this research. As a final application in this book we summarize some of these control theory results. However, unlike our earlier presentations, we give only a brief sample of results with no proofs, with the sole purpose being exposure of readers to another area of science which relies on applied functional analysis in a fundamental way.

One of the earliest (and most distinguished) contributors to use functional analytic techniques to treat control of partial differential equations was J.L. Lions [Li]. There have been many contributions in this spirit in subsequent years; a partial list of interesting references includes [BernHyland, BI2, BaItWa, Curtain84, CurtainSalamon, DaP, GA90, Ito90, KapSal87, KapSal89, Las92, Las95, LLP95, Sch83]. We emphasize only one aspect (feedback control) of the many results available. A feedback control methodology for the second-order infinite dimensional systems discussed in Section 5.1 and Chapter 8 is presented here. These control results are most naturally formulated for first-order systems, and the abstract Cauchy formulation presented in Chapters 7 and 8 for the PDE models of Chapters 8 and 9 provides such a framework. In this Cauchy form, control and approximation results for the finite time and infinite horizon problems can be discussed. All of this is presented in a functional analytic framework which is not only convenient, but, in some aspects, essential for the efficient development of results.

The focus in our presentation will be on unbounded control input operators (the nonhomogeneous terms in the earlier discussions in this book). This is precisely the case, for example, when elements such as piezoceramic or electrostrictive patches are used as control actuators [BSW]. For convenience, the abstract first-order system derived in Section 5.1 and Chapter 8 for structural systems with unbounded input terms is first summarized in an abstract control setting in Section 16.2. Throughout our discussions, emphasis will be on constructing operators and solution semigroups which are compatible with these unbounded control input terms.

Control results for these problems are discussed below. A complete theory exists for the unbounded control input problem in which full state measurements are available. We first present the fundamentals of this theory in the remainder of this chapter. A complete approximation framework for the unbounded linear quadratic regulator (LQR) problem is then summarized in Chapter 17. The theory in both cases utilizes a formulation of the first-order system in terms of sesquilinear forms and corresponding linear operators as developed earlier in this book. The resulting framework provides a rigorous formulation of the infinite dimensional LQR problem with unbounded input term and a summary of the associated convergence theory necessary for approximating the feedback control gains.

As we have noted, results presented here depend upon having knowledge of the full state in order to calculate control gains. In many distributed parameter problems, however, only partial state information is available, and the state must be estimated or reconstructed from observations and a dynamic output feedback law (compensator) developed for controlling the system dynamics. While a theory for compensator design based on sesquilinear forms has been developed for the case of unbounded input/bounded output operators, this theory is much more technically involved and will not be pursued here. Thus our emphasis below is on a careful discussion of infinite dimensional control problems as well as related convergence issues when the LQR and output feedback problems are approximated.

16.2 Abstract Formulation

In Chapters 6 and 7 it was demonstrated that the structural models discussed in Section 6.3 can be written in the weak form

$$\langle \ddot{y}(t), \varphi \rangle_{V^*,V} + \sigma_2(\dot{y}(t), \varphi) + \sigma_1(y(t), \varphi) = \langle Bu(t), \varphi \rangle_{V^*,V} \qquad (16.1)$$

for all $\varphi \in V$ where V is a Hilbert space of test functions with corresponding inner product $\langle \cdot, \cdot \rangle_V$. Here y denotes displacement in a Hilbert space H (state space with inner product $\langle \cdot, \cdot \rangle_H$) and B is a control operator with input u. We have the usual Gelfand triple construction $V \hookrightarrow H \simeq H^* \hookrightarrow V^*$ on the pivot space H with the duality product denoted by $\langle \cdot, \cdot \rangle_{V^*,V}$. The input u is considered in a Hilbert space U. In a typical application involving the control of structural vibrations using piezoceramic actuators, u denotes the voltage to an actuator and $B \in \mathcal{L}(U, V^*)$ is unbounded due to the discontinuous geometry of the patches which leads to external applied line moments in the structure (see [BSW] for further details).

As discussed in earlier chapters, σ_1 incorporates stiffness components of the structural model and σ_2 contains the damping terms (such as Kelvin-Voigt, viscous, or others) for the model. We restrict our discussions here only to the case where both the stiffness and damping forms σ_1 and σ_2 are defined on the same space V and both are V-continuous and V-elliptic. The symmetry, continuity, and coercivity conditions satisfied by the sesquilinear forms are discussed in Chapters 8 and 9, and it was noted that due to the V-continuity (boundedness), one can define operators $A_1, A_2 \in \mathcal{L}(V, V^*)$ by

$$\langle A_i \phi, \psi \rangle_{V^*,V} = \sigma_i(\phi, \psi), \quad i = 1, 2.$$

The weak form (16.1), taken with initial conditions, can then be written equivalently as the second-order system

$$\ddot{y}(t) + A_2 \dot{y}(t) + A_1 y(t) = Bu(t) \quad \text{in } V^*$$

$$y(0) = y_0, \quad y_t(0) = y_1. \qquad (16.2)$$

For consideration of the control problem, it is advantageous to write the system in first-order form. Recall that to this end, we defined the product spaces $\mathcal{H} = V \times H$ and $\mathcal{V} = V \times V$ with the usual product space norms

$$|(\phi_1, \phi_2)|_{\mathcal{H}}^2 = |\phi_1|_V^2 + |\phi_2|_H^2$$

$$|(\phi_1, \phi_2)|_{\mathcal{V}}^2 = |\phi_1|_V^2 + |\phi_2|_V^2.$$

We also recall again that $\mathcal{V} \hookrightarrow \mathcal{H} \simeq \mathcal{H}^* \hookrightarrow \mathcal{V}^*$ forms a Gelfand triple with $\mathcal{V}^* = V \times V^*$ (recall the homework exercise of Section 6.3). The state in $\mathcal{H}$ is denoted by $x(t) = (y(t), \dot{y}(t))$. The stiffness and damping components are combined in the sesquilinear form $\sigma : \mathcal{V} \times \mathcal{V} \to \mathbb{C}$ given by

$$\sigma(\Phi, \Psi) = -\langle \phi_2, \psi_1 \rangle_V + \sigma_1(\phi_1, \psi_2) + \sigma_2(\phi_2, \psi_2)$$

where $\Phi = (\phi_1, \phi_2)$ and $\Psi = (\psi_1, \psi_2)$. Finally, the product space control input term is given by

$$\mathcal{B}u(t) = \begin{bmatrix} 0 \\ Bu(t) \end{bmatrix}.$$

The weak form of the system equations can, as seen in Chapter 6, then be written as the first-order equation

$$\langle \dot{x}(t), \Psi \rangle_{\mathcal{V}^*, \mathcal{V}} + \sigma(x(t), \Psi) = \langle \mathcal{B}u(t), \Psi \rangle_{\mathcal{V}^*, \mathcal{V}} \qquad (16.3)$$

for $\Psi \in \mathcal{V}$. As discussed in Chapter 7, this is formally equivalent to the strong form of the equation in $\mathcal{V}^*$

$$\dot{x}(t) = \mathcal{A}x(t) + \mathcal{B}u(t) \quad \text{in } \mathcal{V}^*$$

$$x(0) = x_0 = \begin{bmatrix} y_0 \\ y_1 \end{bmatrix} \qquad (16.4)$$

where $\mathcal{A}$ is given by

$$\mathcal{D}(\mathcal{A}) = \{(\phi_1, \phi_2) \in \mathcal{H} | \phi_2 \in V, A_1\phi_1 + A_2\phi_2 \in H\}$$

$$\mathcal{A} = \begin{bmatrix} 0 & I \\ -A_1 & -A_2 \end{bmatrix}. \qquad (16.5)$$

We recall (see Section 6.3) that $\mathcal{A}$ is the negative of the restriction to $\mathcal{D}(\mathcal{A})$ of the operator $\tilde{\mathcal{A}} \in \mathcal{L}(\mathcal{V}, \mathcal{V}^*)$ defined by $\sigma(\Phi, \Psi) = \left\langle \tilde{\mathcal{A}}\Phi, \Psi \right\rangle_{\mathcal{V}^*, \mathcal{V}}$ so that $\sigma(\Phi, \Psi) = \langle -\mathcal{A}\Phi, \Psi \rangle_{\mathcal{H}}$ for $\Phi \in \mathcal{D}(\mathcal{A})$, $\Psi \in \mathcal{V}$.

In Section 8.2, it was demonstrated that when σ_2 is V-elliptic (which is the case when a structural model includes strong Kelvin-Voigt damping), the product space sesquilinear form σ is $\mathcal{V}$-elliptic and $\mathcal{A}$ generates an analytic semigroup $\mathcal{T}(t)$ on $\mathcal{V}, \mathcal{H}$ and $\mathcal{V}^*$. (In this case, the $\mathcal{D}(\mathcal{A})$ defined in (16.5) is actually $\mathcal{D}_{\mathcal{H}}(\mathcal{A})$, the domain of $\mathcal{A}$ as a generator in $\mathcal{H}$.) The use of one symbol to denote the semigroups (or general operators) defined on each of the Gelfand triple spaces is common in the literature and should not cause ambiguity.

In weakly damped systems (such as the noise attenuation problem involving the structural acoustic system of [BSW, Chapter 9] in which a vibrating structure is coupled with an adjacent, undamped, or lightly damped acoustic field), however, σ_2 may be only H-semielliptic and σ is *not* V-elliptic. In this case (see Section 8.3), $\mathcal{A}$ generates a semigroup $T(t)$ on $\mathcal{H}$ which is strongly continuous but is *not* analytic. To define a solution compatible with the unbounded input terms in $\{0\} \times V^*$, this semigroup must be extended through extrapolation space techniques to a space which includes the input terms.

For our control systems discussions it is useful to interpret (16.4) in the mild form with a solution given by the variation-of-constants formula

$$x(t) = T(t)x_0 + \int_0^t T(t-s)\mathcal{B}(s)ds \qquad (16.6)$$

where $\mathcal{B}u \in L^2(0, T; V^*)$. The necessity of extrapolating the semigroup to either V^* or a space containing V^* to accommodate the input $\mathcal{B}u$ is now readily apparent and motivates the previously mentioned extensions of Chapter 6 when both σ_1 and σ_2 are assumed V-elliptic. In Chapters 7 and 9, the mild solution (16.6) was shown to be equivalent to the strong and weak solutions when σ_2 is V-elliptic and $\mathcal{B}u$ is sufficiently smooth. Moreover, the well-posedness of the problem (16.4) or equivalently (16.2) was established. Indeed, the well-posedness of (16.6) as a mild solution of (16.4), and equivalence to (16.2), was established under even weaker assumptions on σ_2 (e.g., σ_2 H-semielliptic), but that will not be of concern to us in control since these results typically require more regularity on the input.

16.3 Infinite Dimensional LQR Control: Full State Feedback

We consider the infinite dimensional control problem in the case where full state $(x = (y, \dot{y}))$ information is available for calculation of the feedback control; this is called the *full state feedback* case. Whereas only partial state observations are available in many important distributed parameter applications, the consideration of the full state feedback case is a necessary first step in the development of a state estimator and compensator feedback system. The partial state measurement case will be discussed only briefly in closing Chapter 17.

We also consider observations of the state

$$x_{ob}(t) = \mathcal{C}x(t) \tag{16.7}$$

where $\mathcal{C}$ denotes an observation operator mapping into an observation space Y. When physically implementing the controller, the observation operator is sometimes unbounded (in fact, not well defined) on the state space $\mathcal{H}$ due to the discrete nature of the measurement devices. In this case it may be well defined on $\mathcal{V}$, indeed with $\mathcal{C} \in \mathcal{L}(\mathcal{V}, Y)$. However, the technical details associated with the unbounded operator $\mathcal{C}$ tend to obscure an initial exposition and we will concentrate here on the continuous or bounded operator measurement case $\mathcal{C} \in \mathcal{L}(\mathcal{H}, Y)$. The reader can find various results in [BI2] or [IT] for analysis extending the bounded observation operator results to the unbounded case (see also [Las95]). Again, in all these presentations, functional analysis ideas such as those given in this book comprise essential tools.

Finally, as already noted we restrict our theoretical discussions to the case when σ_2 is V-elliptic and hence σ is $\mathcal{V}$-elliptic and $\mathcal{T}(t)$ is an analytic semigroup. As discussed previously, this includes structural models that incorporate strong damping (e.g., Kelvin-Voigt damping). Although the unbounded control input methodology has been numerically demonstrated as viable and effective for many weakly damped systems, the theory has not yet been completely extended to these systems. To guarantee well-defined trajectories in the three spaces $\mathcal{V}, \mathcal{H}, \mathcal{V}^*$, the following *standing assumptions* are typically made [BI2, IT, PS].

(A1) The semigroup $\mathcal{T}(t)$ is strongly continuous on $\mathcal{H}$ and $\mathcal{V}^*$.

(A2) For every $u(\cdot) \in L_2(0, T; U)$, $\int_0^T \mathcal{T}(T - s)\mathcal{B}u(s)ds \in \mathcal{H}$ and there exists $b > 0$ such that

$$\left| \int_0^T \mathcal{T}(T - s)\mathcal{B}u(s)ds \right|_{\mathcal{H}} \le b\,|u|_{L_2(0,T;U)}\,.$$

(A3) For every $\phi \in \mathcal{H}$ and $\mathcal{C} \in \mathcal{L}(\mathcal{H}, Y)$, there exists $c > 0$ such that

$$|\mathcal{C}\mathcal{T}(t)\phi|_{L_2(0,T;Y)} \le c|\phi|_{\mathcal{V}^*}$$

for all $\phi \in \mathcal{H}$.

(A4) Consider the Hilbert space $\mathcal{W} = \mathcal{D}_{\mathcal{V}^*}(\mathcal{A})$, the domain of $\mathcal{A}$ as an infinitesimal generator in $\mathcal{V}^*$, with the graph norm $|v^*|_{Gr}^2 = |v^*|_{\mathcal{V}^*}^2 + |\mathcal{A}v^*|_{\mathcal{V}^*}^2$ of $\mathcal{A}$ on $\mathcal{V}^*$. Then $\mathcal{W}$ is continuously and densely embedded in $\mathcal{H}$; that is, $\mathcal{W} \hookrightarrow \mathcal{H}$.

We can summarize the role of these assumptions in our control presentation. First, the existence of a strongly continuous semigroup $\mathcal{T}$ on $\mathcal{H}$ follows from the symmetry, continuity, and coercivity properties of the sesquilinear forms σ_1 and σ_2. Furthermore, for the σ_2 V-elliptic case under consideration, the extension to V^*, as required in **(A1)**, follows directly from Theorem 6.4. Moreover, when σ_2 is V-elliptic, **(A2)** follows directly from the equivalence and continuous dependence results of Chapters 7 and 8. For any $u \in L_2(0, T; U)$ and initial value $x_0 \in \mathcal{H}$, this hypothesis implies that the mild solution $x(\cdot)$ defined in Chapter 7 is continuous on $[0, T]$ with values in $\mathcal{H}$. In this case, the output can be expressed as

$$x_{ob}(t) = \mathcal{CT}(t)x_0 + \mathcal{C} \int_0^t \mathcal{T}(t-s)\mathcal{B}u(s)ds \tag{16.8}$$

with well-defined values in Y. If, on the other hand, $x_0 \in V^*$, then x may have values in V^* and **(A3)** must be invoked to obtain a well-defined output by allowing the initial condition component of (16.8) to be continuously extended from $x_0 \in \mathcal{H}$ to $x_0 \in V^*$ so that the bound of **(A3)** holds for all $\phi \in V^*$.

The bound of **(A3)** also follows from standard estimates obtained when $\mathcal{A}$ is defined by (16.5) and σ_2 generating A_2 is V-elliptic (see (3.10) and (6.6) of [BI2]). This property implies that the operator mapping $x_0 \in \mathcal{H}$ to $\mathcal{CT}(\cdot)x_0$ can be extended through continuity to define an L_2 (in time) function $\mathcal{CT}(\cdot)x_0$ for $x_0 \in V^*$. In this manner, the output given by (16.8) can be expressed as a well-defined function in $L_2(0, T; Y)$ for $x_0 \in V^*$.

As noted in [PS], the assumption **(A4)** is not very restrictive and is satisfied in many systems of interest through correct choices of $\mathcal{H}$ and V^*. For the case under consideration in which σ_2 is V-elliptic, the domain of $\mathcal{A}$ defined on V^* satisfies (see (2.9) of [BI2]) $\mathcal{D}_{V^*}(\mathcal{A}) = \{\phi \in V | \mathcal{A}\phi \in V^*\} = V$ so that **(A4)** is readily verified.

16.4 The Finite Horizon Control Problem

For the control problem on a finite time interval $[0, T]$, we consider the quadratic performance index

$$J_T(u, x_0) = \int_0^T \left\{ |\mathcal{C}x(t)|_Y^2 + |\mathcal{R}^{1/2}u(t)|_U^2 \right\} dt + \langle x(T), \mathcal{G}x(T) \rangle_{V^*, V} \tag{16.9}$$

where $\mathcal{G} \in \mathcal{L}(\mathcal{V}^*, \mathcal{V})$ is a nonnegative operator which is self-adjoint in the sense that its restriction to $\mathcal{H}$ is self-adjoint. The positive, self-adjoint operator $\mathcal{R} \in \mathcal{L}(U)$ can be used to weight various components of the control (in the structural applications of [BSW, Chapter 8], it is used to weight the controlling voltage to specific piezoceramic patches). The minimization is performed over $u \in L_2(0, T; U)$ subject to x satisfying (16.4).

The determination of an optimal control (see [BI2, IT]) is facilitated by formulating the output solutions $x_{ob}(t)$ of (16.8) as

$$x_{ob}(t) = \mathcal{M}x_0 + \mathcal{J}u(t)$$

in $L_2(0, T; Y)$. The operators $\mathcal{M} : \mathcal{H} \to L_2(0, T; Y)$ and $\mathcal{J} : L_2(0, T; U) \to L_2(0, T; Y)$ are defined by

$$(\mathcal{M}x_0)(t) = C\mathcal{T}(t)x_0$$

$$(\mathcal{J}u)(t) = C \int_0^t \mathcal{T}(t - s)\mathcal{B}u(s)ds$$

for $t \in (0, T)$. From **(A1)–(A4)**, it follows that $\mathcal{M}$ can be uniquely extended to $\mathcal{V}^*$ so as to be compatible with forces and data in applications. In the spirit of the semigroup operator notation, we will let $\mathcal{M}$ also denote this extension so that $\mathcal{M} \in \mathcal{L}(\mathcal{V}^*, L_2(0, T; Y))$. For fixed terminal time $T > 0$, we also define the operators $\mathcal{J}_T : L_2(0, T; U) \to \mathcal{H}$ and $\mathcal{M}_T : \mathcal{V}^* \to \mathcal{H}$ by

$$\mathcal{J}_T u = \mathcal{G}^{1/2} \int_0^T \mathcal{T}(T - s)\mathcal{B}u(s)ds$$

$$\mathcal{M}_T x_0 = \mathcal{G}^{1/2} \mathcal{T}(T)x_0$$

with $\mathcal{G}^{1/2} \in \mathcal{L}(\mathcal{V}^*, \mathcal{H})$ (further details regarding the characterization $\mathcal{G} = (\mathcal{G}^{1/2})^* \mathcal{G}^{1/2}$ can be found in Corollary 2.1 of [IT]). In terms of these operators, the performance index (16.9) can be written as

$$
\begin{aligned}
J_T(u, x_0) \;=\;& |\mathcal{M}x_0 + \mathcal{J}u|^2_{L_2(0,T;Y)} \\
&+ \; |\mathcal{R}^{1/2}u|^2_{L_2(0,T;U)} + |\mathcal{M}_T x_0 + \mathcal{J}_T u|^2_{\mathcal{H}} \,. \quad (16.10)
\end{aligned}
$$

The optimal control $\bar{u}_T$ is then specified in the following theorem which is Theorem 3.1 in [BI2] or Theorem 2.2 in [IT]; the results also follow from Theorem 2.7 of [PS]. Proofs and further discussions regarding this theorem can be found in these three references.

Theorem 16.1 *In addition to* **(A1)–(A4)**, *assume σ is bounded and $\mathcal{V}$-elliptic and $\mathcal{C} \in \mathcal{L}(\mathcal{H}, Y)$ is a bounded observation operator. For finite T and $x_0 \in \mathcal{H}$, the optimal control $\bar{u}_T$ which minimizes (16.10) is given by*

$$\bar{u}_T = -\left(I + \mathcal{J}^*\mathcal{J} + \mathcal{J}_T^*\mathcal{J}_T\right)^{-1}\left(\mathcal{J}^*\mathcal{M} + \mathcal{J}_T^*\mathcal{M}_T\right)x_0 \ .$$

The performance of the optimal control can be specified in terms of the self-adjoint Riccati operator $\Pi_T \in \mathcal{L}(\mathcal{V}^*, \mathcal{V})$ defined by

$$\Pi_T = (\mathcal{M}^*, \mathcal{M}_T^*)\left(\begin{bmatrix} I & 0 \\ 0 & I \end{bmatrix} + \begin{bmatrix} \mathcal{J} \\ \mathcal{J}_T \end{bmatrix}[\mathcal{J}^*, \mathcal{J}_T^*]\right)^{-1}\begin{pmatrix} \mathcal{M} \\ \mathcal{M}_T \end{pmatrix}$$

which satisfies

$$\langle \Pi_T x_0, x_0 \rangle_{\mathcal{V},\mathcal{V}^*} = J_T(\bar{u}_T, x_0) = \min_u J_T(u, x_0)$$

(we observe that Π_T is self-adjoint in the sense that its restriction to $\mathcal{H}$ is self-adjoint). Moreover, if we let $\Pi_T(t) \equiv \Pi_{T-t}, t \leq T$, then the optimal control is given by the feedback law

$$\bar{u}_T(t) = -\mathcal{R}^{-1}\mathcal{B}^*\Pi_T(t)\bar{x}(t)$$

where $\Pi_T(t)$ satisfies the differential Riccati equation

$$\left(\frac{d}{dt}\Pi_T(t) + \mathcal{A}^*\Pi_T(t) + \Pi_T(t)\mathcal{A} - \Pi_T(t)\mathcal{B}\mathcal{R}^{-1}\mathcal{B}^*\Pi_T(t) + \mathcal{C}^*\mathcal{C}\right)x = 0 \ ,$$

for all $x \in \mathcal{V}$, and $\bar{x}(t)$ denotes the corresponding optimal trajectory.

16.5   The Infinite Horizon Control Problem

The finite horizon control problem stated above is the basis of a number of modern approaches including the so-called *receding horizon* control problem discussed in [DTB] and the references therein. Although control over a specified finite time interval is important in some applications, it is more usual that the time interval is indefinite in length and controls applicable for an unbounded or *infinite horizon* time interval are sought. For these infinite horizon problems, a control u is sought which minimizes the quadratic cost functional

$$J(u, x_0) = \int_0^\infty \left\{|\mathcal{C}x(t)|_Y^2 + |\mathcal{R}^{1/2}u(t)|_U^2\right\} dt \tag{16.11}$$

subject to (16.4). We need several additional concepts to obtain optimal controls for these problems.

Definition (Stab) The pair $(\mathcal{A}, \mathcal{B})$ is said to be *stabilizable* if there exists an operator $\mathcal{K} \in \mathcal{L}(\mathcal{V}^*, U)$ such that $\mathcal{A} - \mathcal{B}\mathcal{K}$ generates an *exponentially stable* semigroup on $\mathcal{V}^*$ (i.e., $|e^{t(\mathcal{A}-\mathcal{B}\mathcal{K})}|_{\mathcal{L}(\mathcal{V}^*)} \leq M e^{-\omega t}$ for $M \geq 1, \omega > 0$).

Definition (Det) The pair $(\mathcal{A}, \mathcal{C})$ is said to be *detectable* if there exists an operator $\mathcal{F} \in \mathcal{L}(Y, \mathcal{V}^*)$ such that $\mathcal{A} - \mathcal{F}\mathcal{C}$ generates an exponentially stable semigroup on $\mathcal{V}^*$.

Under these assumptions and the standing assumptions **(A1)**–**(A4)**, one can find the optimal control in *feedback* form.

Theorem 16.2 *In addition to* **(A1)**–**(A4)**, *assume that* $(\mathcal{A}, \mathcal{B})$ *is stabilizable and* $(\mathcal{A}, \mathcal{C})$ *is detectable. Then the algebraic Riccati equation*

$$\left(\mathcal{A}^*\Pi + \Pi\mathcal{A} - \Pi\mathcal{B}\mathcal{R}^{-1}\mathcal{B}^*\Pi + \mathcal{C}^*\mathcal{C}\right)x = 0 \qquad \text{for all } x \in \mathcal{V}$$

has a unique nonnegative solution $\Pi \in \mathcal{L}(\mathcal{V}^*, \mathcal{V})$, $\mathcal{A} - \mathcal{B}\mathcal{R}^{-1}\mathcal{B}^*\Pi$ *generates an exponentially stable closed loop semigroup* $\mathcal{S}(t) = e^{(\mathcal{A}-\mathcal{B}\mathcal{R}^{-1}\mathcal{B}^*\Pi)t}$ *on* $\mathcal{H}, \mathcal{V}$ *and* $\mathcal{V}^*$, *and the optimal control that minimizes* (16.11) *is given by*

$$\bar{u}(t) = -\mathcal{R}^{-1}\mathcal{B}^*\Pi\bar{x}(t)$$

where $\bar{x}(t) = \mathcal{S}(t)x_0$ *for* $x_0 \in \mathcal{V}^*$.

Remark 3 *The above theorem is essentially Theorem 3.4 in [BI2] or Theorem 2.3 in [IT], and further discussion and proofs can be found therein. A full discussion of analysis leading to this theorem can be found in Theorems 3.3, 3.4 and Remark 3.5 of [PS].*

The conditions of stabilizability and detectability are often referred to as exponential stabilizability *and* exponential detectability *in the literature. We remark that the stabilizability and detectability of the system can be defined in terms of operators* $\mathcal{K} \in \mathcal{L}(\mathcal{H}, U)$ *and* $\mathcal{F} \in \mathcal{L}(Y, \mathcal{H})$ *leading to the generation of exponentially stable semigroups on* $\mathcal{H}$. *One can then use Lemma 3.3 of [BI2] to obtain the desired exponential stability of the semigroups on* $\mathcal{V}$ *and* $\mathcal{V}^*$.

The solution to the algebraic Riccati equation is derived as the limit as $T \to \infty$ of solutions to finite-time integral Riccati solutions given in Theorem 16.1 of Section 16.4.

17 Families of Approximate Control Problems

In the previous chapter we presented control results for the infinite dimensional problem and controls were given in terms of operators and
functions satisfying appropriate smoothness constraints. One cannot of
course implement such operators in practice. Indeed, for these controls
to be implemented, the problem must be discretized and a sequence of
finite dimensional LQR problems considered. As in previous chapters of
this book we consider discretization in the context of Galerkin approximations and approximate solutions are sought in finite dimensional
subspaces $\mathcal{V}^N \subset \mathcal{V} \subset \mathcal{H}$. The bases for these subspaces can consist
of modes, splines, or finite elements which satisfy convergence criteria
to be discussed in this section. We point out that the inclusion of $\mathcal{V}^N$
in $\mathcal{V}$ may be too restrictive for some approximation methods such as
finite differences, and certain spectral and collocation approximations.
In such cases, a relaxation of hypotheses in the manner discussed in
[BKregulator] for the bounded control input analysis can be employed.
We also remark that $\mathcal{V}^N$ is not required to be in $\mathcal{D}_{\mathcal{H}}(\mathcal{A})$. This is important when choosing a basis for $\mathcal{V}^N$ and permits the use of linear splines
in second-order problems and cubic splines in fourth-order systems (see,
e.g., Chapter 12).

We first assume that the approximation method satisfies the following convergence condition (refer to the conditions **(C1)** of Chapter
12):

(CN) For any $x \in \mathcal{V}$, there exists a sequence $\tilde{x}^N \in \mathcal{V}^N$ such that $|x - \tilde{x}^N|_{\mathcal{V}} \to 0$ as $N \to \infty$.

This assumption is standard and is satisfied by most reasonable
approximation methods.

The operator $\mathcal{A}^N : \mathcal{V}^N \to \mathcal{V}^N$ which approximates $\mathcal{A}$ is defined by
restricting σ to $\mathcal{V}^N \times \mathcal{V}^N$; hence (again see Chapter 12)

$$\langle -\mathcal{A}^N \Phi, \Psi \rangle_{\mathcal{H}} = \sigma(\Phi, \Psi) \qquad \text{for all } \Phi, \Psi \in \mathcal{V}^N .$$

For each N, the C_0 semigroup on $\mathcal{V}^N$ that is generated by $\mathcal{A}^N$ is denoted
by $\mathcal{T}^N(t)$. The control operator $\mathcal{B} \in \mathcal{L}(U, \mathcal{V}^*)$ is approximated by
$\mathcal{B}^N \in \mathcal{L}(U, \mathcal{V}^N)$ which is defined through duality by

$$\langle \mathcal{B}^N u, \Psi \rangle_{\mathcal{H}} = \langle u, \mathcal{B}^* \Psi \rangle_U \qquad \text{for all } u \in U , \ \Psi \in \mathcal{V}^N . \tag{17.1}$$

The observation operator $\mathcal{C}^N$ is simply obtained by restricting $\mathcal{C}$ to $\mathcal{V}^N$. Finally, we let P^N denote the usual orthogonal projection of $\mathcal{H}$ onto $\mathcal{V}^N$ which by definition satisfies

(i) $P^N \Phi \in \mathcal{V}^N$ for $\Phi \in \mathcal{H}$

(ii) $\langle P^N \Phi - \Phi, \Psi \rangle_{\mathcal{H}} = 0$ for all $\Psi \in \mathcal{V}^N$.

This projection can be extended to $P^N \in \mathcal{L}(\mathcal{V}^*, \mathcal{V}^N)$ by replacing the $\mathcal{H}$-inner product $\langle \Phi, \Psi \rangle_{\mathcal{H}}$ in the definition (ii) by $\langle \Phi, \Psi \rangle_{\mathcal{V}^*, \mathcal{V}}$ and considering $\Phi \in \mathcal{V}^*$. The approximate problem corresponding to (16.3) can then be formulated as

$$\frac{d}{dt} \langle x^N(t), \Psi \rangle_{\mathcal{H}} + \sigma(x^N(t), \Psi) = \langle \mathcal{B}^N u(t), \Psi \rangle_{\mathcal{H}} \qquad \text{for all } \Psi \in \mathcal{V}^N$$

$$x^N(0) = P^N x_0,$$

$$(17.2)$$

with the solution

$$x^N(t) = \mathcal{T}^N(t) P^N x_0 + \int_0^t \mathcal{T}^N(t-s) \mathcal{B}^N u(s) ds \ . \qquad (17.3)$$

If **(CN)** is satisfied, x denotes the solution to (16.3), and $e^N(t) \equiv x^N(t) - x(t)$ denotes the error, then one might expect the convergence

$$\left| e^N(t) \right|_{\mathcal{H}} \to 0$$

$$\int_0^t \left| e^N(s) \right|_{\mathcal{V}}^2 ds \to 0$$

as $N \to \infty$. Indeed this is true and one can see, for example, Chapter III in [Li] for arguments for such results.

17.1 The Finite Horizon Problem: Approximate Control Gains

In a previous section we considered the minimization of the functional (16.9) to obtain the optimal control over a finite time interval $(0, T)$. The corresponding N^{th} approximate problem in $\mathcal{V}^N$ concerns the minimization of the cost functional

$$J_T^N(u, x_0) = \int_0^T \left\{ \left| \mathcal{C}^N x^N(t) \right|_Y^2 + \left| \mathcal{R}^{1/2} u(t) \right|_U^2 \right\} dt + \langle \mathcal{G}^N x^N(T), x^N(T) \rangle_{\mathcal{H}}$$

$$(17.4)$$

subject to

$$\frac{d}{dt}x^N(t) = \mathcal{A}^N x^N(t) + \mathcal{B}^N u(t), \quad 0 < t < T$$

$$x^N(0) = P^N x_0 .$$

(17.5)

The symmetric operator $\mathcal{G}^N$ is defined by $\mathcal{G}^N = P^N \mathcal{G} P^N$.

Since the trajectories of (17.5) evolve in $\mathcal{V}^N$, finite dimensional control theory can be used to determine the optimal control which minimizes (17.4). The main question of interest then concerns the convergence of the corresponding finite dimensional Riccati operators and approximate controls to their infinite dimensional counterparts. The desired convergence properties of the controls and Riccati operators for the finite-time problem can be found in [BI2, IT] and are summarized in the following theorem.

Theorem 17.1 *Assume* **(A1)–(A4)** *and suppose* $\mathcal{A}$ *is defined by (16.5) and* $\mathcal{B} \in \mathcal{L}(U, \mathcal{V}^*), \mathcal{C} \in \mathcal{L}(\mathcal{V}, Y), \mathcal{G} \in \mathcal{L}(\mathcal{V}^*, \mathcal{V})$ *are the previously discussed control, observation, and terminal weight operators. Consider an approximation scheme satisfying* **(CN)** *with the corresponding approximate operators* $\mathcal{A}^N, \mathcal{B}^N, \mathcal{C}^N, \mathcal{G}^N$. *Moreover, let* $\Pi_T^N(t), t \leq T$ *denote the solution to the Riccati equation*

$$\frac{d}{dt}\Pi_T^N(t) + \mathcal{A}^{N*}\Pi_T^N(t) + \Pi_T^N(t)\mathcal{A}^N - \Pi_T^N(t)\mathcal{B}^N \mathcal{R}^{-1}\mathcal{B}^{N*}\Pi_T^N(t)$$
$$+ \mathcal{C}^{N*}\mathcal{C}^N = 0$$

$$\Pi_T^N(T) = P^N \mathcal{G} P^N = \mathcal{G}^N$$

in $\mathcal{V}^N$, *and let* $\bar{u}_T^N$ *denote the optimal control for the* N^{th} *approximate problem (17.4)-(17.5). The following convergence is then obtained:*

(i) $\left|\bar{u}_T^N - \bar{u}_T\right|_{L_2(0,T;U)} \to 0$ *for all* $x_0 \in \mathcal{H}$

(ii) $\left|\Pi_T^N(t)P^N x - \Pi_T(t)x\right|_{\mathcal{H}} \to 0$ *uniformly in* $t \in [0, T]$ *for all* $x \in \mathcal{H}$.

The optimal control $\bar{u}_T^N$ *to the* N^{th} *approximate problem is given by*

$$\bar{u}_T^N(t) = -\mathcal{R}^{-1}\mathcal{B}^{N*}\Pi_T^N(t)x^N(t) .$$

Furthermore, if $\mathcal{C}$ *is bounded (i.e.,* $\mathcal{C} \in \mathcal{L}(\mathcal{H}, Y)$) *and* **(A3)** *is satisfied, then (ii) can be strengthened to yield convergence in the uniform operator topology, i.e.,*

(ii') $\Pi_T^N(t)$ *converges to* $\Pi_T(t)$ *in* $\mathcal{L}(\mathcal{V}^*, \mathcal{V})$ *uniformly for* t *in* $[0, T]$.

We refer the reader to [BI2], Theorems 4.5 and 4.6 (as well as Remark 3.2(1), Lemma 6.1, and the discussions following the lemma) and [IT], Theorem 2.4 for further details and proofs.

17.2 The Infinite Horizon Problem: Approximate Control Gains

We turn next to approximate problems over an indefinite horizon, that is, $T = \infty$. The optimization problem in this case consists of finding $u \in L_2(0, \infty; U)$ which minimizes

$$J^N(u, x_0) = \int_0^\infty \left\{ \left| \mathcal{C}^N x^N(t) \right|_Y^2 + \left| \mathcal{R}^{1/2} u(t) \right|_U^2 \right\} dt \qquad (17.6)$$

subject to x^N satisfying the evolution equation

$$\frac{d}{dt} x^N(t) = \mathcal{A}^N x^N(t) + \mathcal{B}^N u(t) \quad , \quad t > 0 \qquad (17.7)$$
$$x^N(0) = P^N x_0 \ .$$

In general, this problem involves the bounded approximate control operators $\mathcal{B}^N$ which should converge to an unbounded operator $\mathcal{B}$. For the discussion here, we will again assume that full state measurements are available so that $\mathcal{C}^N$ is the restriction of $\mathcal{C} \in \mathcal{L}(\mathcal{H}, Y)$ to $\mathcal{V}^N$. The assumption of a bounded observation operator simplifies the presentation here.

On considering the functional J^N of (17.6) we see that two limits are involved in the convergence process, namely $N \to \infty$ and $T \to \infty$. As detailed in the proof of Theorem 2.2 of [BKregulator] for the bounded control operator and Theorem 2.6 of [IT] for an unbounded operator, convergence arguments typically start with the consideration of the finite-time minimization problem

$$\text{minimize} \int_0^T \left\{ \left| \mathcal{C}^N x^N(t) \right|_Y^2 + \left| \mathcal{R}^{1/2} u(t) \right|_U^2 \right\} dt$$

subject to (17.7).

The strategy is to then bound the observed trajectory and control as T increases in a manner which guarantees the existence of nonnegative self-adjoint solutions $\Pi^N \in \mathcal{L}(\mathcal{V}^*, \mathcal{V})$ to the Nth approximate algebraic Riccati equation in $\mathcal{V}^N$

$$\mathcal{A}^{N^*}\Pi^N + \Pi^N \mathcal{A}^N - \Pi^N \mathcal{B}^N \mathcal{R}^{-1} \mathcal{B}^{N^*}\Pi^N + \mathcal{C}^{N^*}\mathcal{C}^N = 0 \qquad (17.8)$$

for N sufficiently large.

We recall from the discussions above of the corresponding infinite dimensional problems that the condition of stabilizability of $(\mathcal{A}, \mathcal{B})$ was used to guarantee the existence of a Riccati solution $\Pi \in \mathcal{L}(\mathcal{V}^*, \mathcal{V})$ to the algebraic Riccati equation of Theorem 16.2. The uniqueness of the solution and exponential stability of the closed loop semigroup were obtained using the hypothesis of detectability of $(\mathcal{A}, \mathcal{C})$. As discussed in [BI2] and [IT], the analogous concepts of uniform stabilizability of $(\mathcal{A}^N, \mathcal{B}^N)$ and uniform detectability of $(\mathcal{A}^N, \mathcal{C}^N)$ can be used to prove the existence of a unique Riccati solution Π^N and exponential stability of the closed loop semigroup $\mathcal{S}^N(t)$. We give the formal definitions needed.

Definition (UnifStab) The pair $(\mathcal{A}^N, \mathcal{B}^N)$ is said to be *uniformly stabilizable* if there exist constants $M_1 \geq 1, \omega_1 > 0$ independent of N and a sequence of operators $\mathcal{K}^N \in \mathcal{L}(\mathcal{V}^N, U)$ such that $\sup_N |\mathcal{K}^N| < \infty$ and

$$\left| e^{t(\mathcal{A}^N - \mathcal{B}^N \mathcal{K}^N)} P^N x \right|_{\mathcal{H}} \leq M_1 e^{-\omega_1 t} |x|_{\mathcal{H}}$$

for $x \in \mathcal{H}$.

Definition (UnifDet) The pair $(\mathcal{A}^N, \mathcal{C}^N)$ is said to be *uniformly detectable* if there exist constants $M_2 \geq 1, \omega_2 > 0$ independent of N and a sequence of operators $\mathcal{F}^N \in \mathcal{L}(Y, \mathcal{V}^N)$ such that $\sup_N |\mathcal{F}^N| < \infty$ and

$$\left| e^{t(\mathcal{A}^N - \mathcal{F}^N \mathcal{C}^N)} P^N x \right|_{\mathcal{H}} \leq M_2 e^{-\omega_2 t} |x|_{\mathcal{H}}$$

for $x \in \mathcal{H}$.

To treat $x \in \mathcal{V}^*$, the projections can be extended to $P^N \in \mathcal{L}(\mathcal{V}^*, \mathcal{V}^N)$ with norms altered accordingly.

Precisely stated convergence results for the infinite horizon problem can then be summarized in the following theorem. The proof can be obtained from the discussions of Theorem 4.8 and Section 6.1 in [BI2] or Theorem 2.6 in [IT] along with the observation that for second-order systems, the operator $\mathcal{B}$ has the form $\mathcal{B} = (0, B)^T$. To obtain the desired result, the following assumption is also required (note we do *not* have $i : V \hookrightarrow H$ compact!).

(A5) The injection $i : V \hookrightarrow H$ is compact.

Theorem 17.2 *Suppose* **(A1)**–**(A5)** *hold. Let the sesquilinear form σ associated with the first-order system (16.3) be V-continuous and V-elliptic. Assume that the operators $\mathcal{A}, \mathcal{B}, \mathcal{C}$ of (16.4), (16.7) satisfy: $(\mathcal{A}, \mathcal{B})$ is stabilizable and $(\mathcal{A}, \mathcal{C})$ is detectable where $\mathcal{B} \in \mathcal{L}(U, V^*)$ is unbounded and $\mathcal{C} \in \mathcal{L}(H, Y)$ is bounded. Assume an approximation method which satisfies* **(CN)**. *Finally, suppose that for fixed N_0 and $N > N_0$, the pair $(\mathcal{A}^N, \mathcal{B}^N)$ is uniformly stabilizable and $(\mathcal{A}^N, \mathcal{C}^N)$ is uniformly detectable.*

Then for N sufficiently large, there exists a unique nonnegative self-adjoint solution $\Pi^N \in \mathcal{L}(V^*, V)$ to the Nth approximate algebraic Riccati equation (17.8) in V^N. There also exist constants $M_3 \geq 1$ and $\omega_3 > 0$ independent of N such that $\mathcal{S}^N(t) = e^{(\mathcal{A}^N - \mathcal{B}^N \mathcal{R}^{-1} \mathcal{B}^{N^*} \Pi^N)t}$ satisfies

$$\left| \mathcal{S}^N(t) \right|_{V^N} \leq M_3 e^{-\omega_3 t} \quad , \quad t > 0$$

or equivalently

$$\left| e^{t(\mathcal{A}^N - \mathcal{B}^N \mathcal{R}^{-1} \mathcal{B}^{N^*} \Pi^N)} P^N x_0 \right|_H \leq M_3 e^{-\omega_3 t} |x_0|_H \quad , \quad t > 0 \ , \ x_0 \in \mathcal{H} \ .$$

Additionally, the convergences

$$\Pi^N P^N x \overset{strongly}{\rightarrow} \Pi x \quad in \ V \ \ for \ every \ x \in V^*$$

$$\left| \mathcal{B}^{N^*} \Pi^N P^N - \mathcal{B}^* \Pi \right|_{\mathcal{L}(\mathcal{H}, U)} \rightarrow 0 \ ,$$

as $N \rightarrow \infty$, of the Riccati and control operators are obtained. Moreover, the feedback system operator $\mathcal{A} - \mathcal{B}\mathcal{R}^{-1}\mathcal{B}^{N^*}\Pi^N$ generates an exponentially stable analytic semigroup on $\mathcal{H}$ and for every $x_0 \in \mathcal{H}$,

$$J\left(-\mathcal{B}^{N^*} \Pi^N x(\cdot), x_0 \right) - J(\overline{u}, x_0) \leq \varepsilon(N) |x_0|_{\mathcal{H}}^2$$

where $\varepsilon(N) \rightarrow 0$ as $N \rightarrow \infty$.

The final condition of Theorem 17.2 implies that the finite dimensional control yields an exponentially stable semigroup when applied to the original infinite dimensional system. This is of *great practical importance* since it represents the case when *computed controls are applied to the actual physical systems of interest.*

Theorem 17.2 is the second-order system analog of Theorem 4.8 of [BI2], which is for first-order systems. To obtain Theorem 17.2 directly from Theorem 4.8 in [BI2], one would require that the embedding $\mathcal{V} \hookrightarrow \mathcal{H}$ be compact. This is not true for our second-order systems where $\mathcal{V} = V \times V$ and $\mathcal{H} = V \times H$. However, one can use arguments similar to those employed in Theorem 4.8 of [BI2] along with **(A5)** and the special structure of our second-order systems written in first-order form to prove Theorem 17.2.

For the results here, the system is assumed to be strongly damped or essentially parabolic in nature; hence σ_2 is V-elliptic and σ is $\mathcal{V}$-elliptic. As discussed in Lemmas 4.2 and 4.3 of [BI2], this leads to resultant estimates for $\mathcal{A}^N$ and bounds and estimates for the open loop semigroups $\mathcal{T}(t)$ and $\mathcal{T}^*(t)$ in both the $\mathcal{H}$ and $\mathcal{V}$ norms and hence *convergence of the approximating and adjoint approximating semigroups* $\mathcal{T}^N$ and $\mathcal{T}^{N^*}$ (see (i)-(iii) of Lemma 4.3 in [BI2]). Moreover, the convergence results summarized in Lemma 4.4 of [BI2] for $\mathcal{M}^N$ and $\mathcal{M}^{N^*}$ are typical of the consistency criteria that can be obtained. Conditions such as the adjoint convergence $\mathcal{T}^{N^*}(t)P^N x \to \mathcal{T}^*(t)x$, as $N \to \infty$, may appear on first encounter as purely mathematical with little physical relevance to the actual problem. However, as demonstrated by examples in [BBu2, BKap1, BRI, KapSal87, KapSal89] in which the calculation of feedback gains for a delay equation are considered, failure of approximation methods to satisfy adjoint convergence can lead to *gains which do not strongly converge.*

As a final comment on these results, we note that care must also be taken when choosing an approximation method to guarantee that the uniform stabilizability margins are preserved under approximation. As illustrated in the context of delay equations [BKap1] and weakly damped PDE systems [BaItWa], approximation methods which are quite adequate for open loop simulations and/or parameter estimation (e.g., finite differences and finite elements) may not preserve these margins. This can slow or even prevent convergence of the feedback gains and illustrates the care that must be taken when choosing an approximation methodology that is suitable for control calculations.

The hypotheses of uniform stabilizability and uniform detectability of $(\mathcal{A}^N, \mathcal{B}^N)$ and $(\mathcal{A}^N, \mathcal{C}^N)$, respectively, of Theorem 17.2 yield the desired convergence results. However, these conditions are often difficult to verify directly, and we now discuss more readily checked assumptions which yield the uniform stabilizability and detectability conditions.

For general first-order systems, uniform stabilizability is obtained using the following lemma which is Lemma 4.7 in [BI2] and is the unbounded input analogue of the *preservation of exponential stabilizablity (POES)* condition first introduced in [BKregulator].

Lemma 17.1 *Suppose $(\mathcal{A}, \mathcal{B})$ is stabilizable and the injection $\mathcal{V} \hookrightarrow \mathcal{H}$ is compact. Then there exists a positive integer N_0 such that for all $N > N_0$, the pair $(\mathcal{A}^N, \mathcal{B}^N)$ is uniformly stabilizable.*

For truly first-order systems ($\mathcal{H} = H$, $\mathcal{V} = V$, as opposed to $\mathcal{H} = V \times H$, $\mathcal{V} = V \times V$), the compact injection condition is quite often satisfied through natural choices for the state and test function space (e.g., $\mathcal{H} = H = L_2(\Omega), \mathcal{V} = V = H^1(\Omega)$) due to standard Sobolev embedding theorems [Adams]. In such cases, Lemma 17.1 can be combined with the first-order analogue of Theorem 17.2 to yield the desired convergence results with hypotheses that can be verified in physical applications.

On the other hand, in second-order systems that are written in first-order form through a product space formulation, the embedding $i : \mathcal{V} \hookrightarrow \mathcal{H}$ is definitely *not* compact and hence Lemma 17.1 is not useful. Lemma 17.2 below contains conditions used (along with Theorem 17.2) to obtain uniform exponential bounds on the approximating semigroups $S^N(t)$ on $\mathcal{H}^N \subset \mathcal{H} = V \times H$ for the second-order systems of interest to us.

Lemma 17.2 *Assume that **(A5)** holds. Moreover, suppose that the damping sesquilinear form can be decomposed as $\sigma_2 = \delta\sigma_1 + \tilde{\sigma}_2$, for some $\delta > 0$, where the continuous sesquilinear form $\tilde{\sigma}_2$ satisfies for some real constant μ*

$$Re\, \tilde{\sigma}_2(\phi, \phi) \geq -\frac{\delta}{2}|\phi|_V^2 - \mu|\phi|_H^2 \quad \text{for all } \phi \in V .$$

Finally, suppose that the operator $A_1^{-1}\widetilde{A}_2$, *where* $\widetilde{A}_2 \in \mathcal{L}(V, V^*)$ *is defined by* $\left\langle \widetilde{A}_2\phi, \eta \right\rangle_{V^*,V} = \tilde{\sigma}_2(\phi, \eta)$, *is compact on* V.

Let $\mathcal{T}$ *denote the open loop semigroup generated by the product space operator* $\mathcal{A}$ *and let* $\mathcal{T}^N$ *be generated by* $\mathcal{A}^N$. *If for some* $\omega \in \mathbb{R}$ *and* $M \geq 1$

$$|\mathcal{T}(t)|_{\mathcal{L}(\mathcal{H})} \leq Me^{\omega t} \quad , \quad t \geq 0 ,$$

then for any $\varepsilon > 0$ *there exists an integer* N_ε *such that for* $N \geq N_\varepsilon$

$$|\mathcal{T}^N(t)P^N|_{\mathcal{L}(\mathcal{H})} \leq \widetilde{M}e^{(\omega+\varepsilon)t} \quad , \quad t \geq 0$$

for some constant $\widetilde{M} > 0$ *independent of* N.

This is Lemma 6.2 in [BI2] and the proof can be found there. When combined with Theorem 17.2, one then obtains convergence of the Riccati and control operators for those special first-order systems corresponding to second-order problems.

To apply the lemma, one must first verify that for the system under consideration, the operator $A_1^{-1}\widetilde{A}_2$ is compact in V. In considering this proposition, it can be advantageous to use the property that the product of a compact linear operator and a bounded linear operator yields a compact linear operator (see, for example, Lemma 8.3-2 of [Kreyszig]). For many problems of interest, A_1 is a differential operator whose inverse is compact, whereas $\widetilde{A}_2$ is quite often bounded. This is illustrated in the following example in the context of the Euler-Bernoulli beam model from earlier sections discussed once again in a control setting.

17.2.1   Example 6 Again!

We consider again the cantilever Euler-Bernoulli beam having length ℓ with fixed end at $\xi = 0$ given in Section 5.1. We assume Kelvin-Voigt internal damping and viscous or air external damping. Control is assumed to be applied through a pair of identical piezoceramic patches as discussed in Chapter 4 of [BSW].

The density, Young's modulus, moment of inertia, viscous damping, and Kelvin-Voigt damping coefficients are denoted by ρ, E, I, γ, and c_D, respectively, and are assumed constant. This assumption would be reasonable for the case of a homogeneous beam or a beam with piezo-ceramic patches embedded in a manner so that material properties do not vary across the region of the patches. We again consider only the transverse displacement which is denoted by y. An external moment is applied through an applied voltage $u(t)$ to a pair of embedded piezo-ceramic patches operated out-of-phase. As discussed in [BSW], the equation describing the transverse beam motion is

$$\rho\frac{\partial^2 y}{\partial t^2} + \gamma\frac{\partial y}{\partial t} + \frac{\partial^2}{\partial \xi^2}(M_\xi) = \frac{\partial^2}{\partial \xi^2}\left(\mathcal{K}^B u(t)\chi_{pe}(\xi)\right)$$

$$y(t,0) = \frac{\partial}{\partial \xi}y(t,0) = 0 \ , \ M_\xi(t,\ell) = \frac{\partial}{\partial \xi}(M_\xi)(t,\ell) = 0$$

where the derivatives $\frac{\partial}{\partial \xi}, \frac{\partial^2}{\partial \xi^2}$ are taken in the distributional sense and the internal moment M_ξ is given as in Section 5.1. We note that $\mathcal{K}^B$ is a patch constant while $\chi_{pe}(\xi)$ denotes the characteristic function over the region containing the patch pair.

As before we choose state space $H = L_2(0,\ell)$, and the space of test functions $V = H_L^2(0,\ell)$, each taken with their usual inner products. For these choices, the compact embedding of V into H follows immediately from standard Sobolev theory [Adams]. Recall (see Chapter 8) that the equation of motion can then be written in the weak form

$$\langle\ddot{y}(t),\varphi\rangle_{V^*,V} + \sigma_2(\dot{y}(t),\varphi) + \sigma_1(y(t),\varphi) = \langle Bu(t),\varphi\rangle_{V^*,V} \quad \text{for all } \varphi \in V$$

where

$$\sigma_1(\varphi,\psi) = \int_0^\ell EI\varphi''\psi''dy$$

$$\sigma_2(\varphi,\psi) = \int_0^\ell c_D I\varphi''\psi''dy + \gamma\int_0^\ell \varphi\psi dy$$

are continuous, V-elliptic, and symmetric, and $Bu(t)$ is an unbounded (V^* valued) input operator. Thus the results of Chapters 7 and 8 concerning variational solutions and semigroup generation hold for this example. The first-order form of the equation along with the associated generator $\mathcal{A}$ are precisely defined as in (16.4), (16.5).

For the constant coefficient case under consideration here, the damping form σ_2 can be decomposed as $\sigma_2 = \delta\sigma_1 + \tilde{\sigma}_2$ where δ and $\tilde{\sigma}_2$ are given by $\delta = \frac{c_D}{E}$ and $\tilde{\sigma}_2(\varphi,\psi) = \gamma\int_0^\ell \varphi\psi dy$. Clearly, $\tilde{\sigma}_2$ satisfies

$$Re\ \tilde{\sigma}_2(\varphi,\varphi) = \frac{\gamma}{\rho}|\varphi|_H^2 \geq -\frac{\delta}{2}|\varphi|_V^2$$

for all $\varphi \in V$. The operator $\tilde{A}_2$ generated by $\tilde{\sigma}_2$ is given by $\tilde{A}_2 = \gamma I$ and is therefore bounded. Since the injections $i : V \hookrightarrow H$, $i^* : H \hookrightarrow V^*$ are compact, we have that $A_1^{-1} \in \mathcal{L}(V^*, V)$ can be written as an operator on $V \to V$ by $A_1^{-1} = A_1^{-1} i^* i$. Thus A_1^{-1} is compact on V and hence $A_1^{-1} \tilde{A}_2$ is compact on V.

Finally, the exponential stability of the open loop semigroup $\mathcal{T}(t)$ generated by $\mathcal{A}$ is guaranteed by the following theorem which is discussed in [BI]:

Theorem 17.3 *Suppose σ_1 is V-elliptic, V-continuous, and symmetric and σ_2 is H-elliptic, continuous, and symmetric. Then $\mathcal{A}$ defined via (16.5) is the infinitesimal generator of a C_0-semigroup $\mathcal{T}(t)$ in $\mathcal{H} = V \times H$ that is exponentially stable (i.e., $|\mathcal{T}(t)z_0|_{\mathcal{H}} \leq M e^{-\omega t} |z_0|_{\mathcal{H}}$).*

From Theorem 17.3, it also follows that $(\mathcal{A}, \mathcal{B})$ is stabilizable and $(\mathcal{A}, \mathcal{C})$ is detectable for the example under discussion.

The hypotheses for Lemma 17.2 are satisfied for Example 6 and one therefore obtains uniform bounds on the approximating semigroups. When combined with the conclusions of Theorem 17.2, this then leads to the convergence of the Riccati and control operators in this example. We remark that these arguments are not unique to the Euler-Bernoulli beam model and other structural models (e.g., the plate models discussed in Chapters 5 and 8 of [BSW]) can be readily analyzed in a similar manner with the functional analytic tools presented in this book.

The convergence of Theorem 17.2 is obtained under the assumption of a bounded observation operator $\mathcal{C} \in \mathcal{L}(\mathcal{H}, Y)$. As noted previously, however, the observation operator in many current applications is unbounded (discontinuous) due to the spatially discrete nature of commonly used measurement devices (e.g., accelerometers, laser vibrometers, proximity sensors, piezoelectric sensors, strain gauges, microphones). The reader is referred to Sections 5 and 6.2 of [BI2] for an extension of the theory for Theorem 17.2 to the *unbounded observation operator* case, $\mathcal{C} \in \mathcal{L}(V, Y)$. The arguments for this case are somewhat more technical in that they involve intermediate spaces, but theoretical results quite similar to those outlined above for bounded observation operators can be obtained for the unbounded observation case. The theory concerning the form of the observation operator takes on practical significance when designing an estimator or observer for *state reconstruction* from *partial state measurements*, and further discussion concerning this topic can be found in [BSW]. We remark that in this case one develops an operator theory version of dynamic compensators or *Kalman filters* [GA90] and a "separation principle" which allows one to solve separately an LQR problem (as discussed above) and a dual *"state estimator" problem.* Details in a functional analytic setting rely on many of the tools we have developed here and are summarized in [BSW].

The analysis outlined in the previous sections was developed in the context of a sesquilinear functional formulation of the control problem. This provided a natural framework in which to incorporate the unbounded input and output operators. Other functional analysis approaches involving primarily operator-based theory have also been developed for analysis of infinite dimensional LQR problems with unbounded input and output operators, and the reader is referred to [LLP95] and references therein for development of theory in the context of this latter methodology.

References

[Remark] In many of the references below, we refer to CRSC-TRXX-YY. These refer to early Technical Report versions of manuscripts which can be found on the Center for Research in Scientific Computation website at North Carolina State University where XX refers to the year, e.g., $XX = 03$ is 2003, $XX = 99$ is 1999, while the YY refers to the number of the report in that year. These can be found at and downloaded from http://www.ncsu.edu/crsc/reports.html where they are listed by year.

[Adams] R. A. Adams, *Sobolev Spaces*, Academic Press, New York, 1975.

[Allen] L. J. S. Allen, *An Introduction to Stochastic Processes with Applications to Biology*, Prentice Hall, New Jersey, 2003.

[Amann] H. Amann, et al., (eds.), *Functional Analysis and Evolution Equations: The Günter Lumer Volume*, Birkhäuser, Basel, 2008.

[AD] A. C. Atkinson and A. N. Donev, *Optimum Experimental Designs*, Oxford University Press, New York, 1992.

[Aubin] J. P. Aubin, *Optima and Equilbria: An Introduction to Nonlinear Analysis*, Springer-Verlag, Berlin, 1993.

[Balanis] C.A. Balanis, *Advanced Engineering Electromagnetics*, John Wiley & Sons, New York, 1989.

[BJR] A. Bamberger, P. Joly, and J. E. Roberts, Second-order absorbing boundary conditions for the wave equation: a solution for the corner problem, *SIAM Journal on Numerical Analysis,* **27** (1990), 323–352.

[Banks79] H. T. Banks, Approximation of nonlinear functional differential equation control systems, *Journal of Optimization Theory and Applications,* **29** (1979), 383–408.

[Banks82] H. T. Banks, Identification of nonlinear delay systems using spline methods, in: V. Lakshmikantham (ed.), *Nonlinear Phenomena in Mathematical Sciences*, Academic Press, New York, 1982, pp. 47–55.

[BBDS1] H. T. Banks, J. E. Banks, L. K. Dick, and J. D. Stark, Estimation of dynamic rate parameters in insect populations undergoing sublethal exposure to pesticides, CRSC-TR05-22, NCSU, May, 2005; *Bulletin of Mathematical Biology*, **69** (2007), 2139–2180.

[BBDS2] H. T. Banks, J. E. Banks, L. K. Dick, and J. D. Stark, Time-varying vital rates in ecotoxicology: selective pesticides and aphid population dynamics, *Ecological Modelling*, **210** (2008), 155–160.

[BBJ] H. T. Banks, J. E. Banks, and S. L. Joyner, Estimation in time-delay modeling of insecticide-induced mortality, CRSC-TR08-15, October, 2008; *Journal of Inverse and Ill-posed Problems*, **17** (2009), 101–125.

[BaBar] H. T. Banks and J. M. Bardsley, Wellposedness for systems arising in time domain electromagnetics in dielectrics, CRSC-TR03-27, July, 2003; *International Journal of Pure and Applied Mathematics*, **46** (2008), 1–18.

[BBi] H. T. Banks and K. L. Bihari, Modeling and estimating uncertainty in parameter estimation, *Inverse Problems*, **17** (2001), 1–17.

[BBi2] H. T. Banks and K. L. Bihari, Analysis of thermal conductivity in composite adhesives, CRSC-TR01-20, August, 2001; *Numerical Functional Analysis and Optimization*, **23** (2002), 705–745.

[BBo] H. T. Banks and V. A. Bokil, Parameter identification for dispersive dielectrics using pulsed microwave interrogating signals and acoustic wave induced reflections in two and three dimensions, CRSC-TR04-27, July, 2004; Revised version appeared as: A computational and statistical framework for multidimensional domain acoustoptic material interrogation, *Quarterly of Applied Mathematics*, **63** (2005), 156–200.

[BaBoHetalShrimp] H. T. Banks, V. A. Bokil, S. Hu, F. C. T. Allnutt, R. Bullis, A. K. Dhar, and C. L. Browdy, Shrimp biomass and viral infection for production of biological countermeasures, CRSC-TR05-45, December, 2005; *Mathematical Biosciences and Engineering*, **3** (2006), 635–660.

[BanksBortzIP] H. T. Banks and D. M. Bortz, Inverse problems for a class of measure dependent dynamical systems, CRSC-TR04-33,

September, 2004; *Journal of Inverse and Ill-posed Problems*, **13** (2005), 103–121.

[BBH] H. T. Banks, D. M. Bortz, and S. E. Holte, Incorporation of variability into the mathematical modeling of viral delays in HIV infection dynamics, *Mathematical Biosciences*, **183** (2003), 63–91.

[BBPP] H. T. Banks, D. M. Bortz, G. A. Pinter, and L. K. Potter, Modeling and imaging techniques with potential for application in bioterrorism, CRSC-TR03-02, January, 2003; Chapter 6 in *Bioterrorism: Mathematical Modeling Applications in Homeland Security*, (H.T. Banks and C. Castillo-Chavez, eds.), Frontiers in Applied Math, **FR28**, SIAM, Philadelphia, PA, 2003, 129–154.

[BBKW] H. T. Banks, L. W. Botsford, F. Kappel, and C. Wang, Modeling and estimation in size structured population models, LCDS-CCS Report 87-13, Brown University; *Proceedings of 2nd Course on Mathematical Ecology*, (Trieste, December 8-12, 1986) World Press (1988), Singapore, 521–541.

[BBCFUVW] H. T. Banks, B. Boudreaux, A. K. Criner, K. Foster, C. Uttal, T. Vogel, and W. P. Winfree, Thermal interrogation of porous materials, CRSC-TR08-11, September, 2008; short version: Thermal based damage detection in porous materials, *Inverse Problems in Science and Engineering*, **18** (2009), 835–851.

[BBL] H. T. Banks, M. W. Buksas, and T. Lin. *Electromagnetic Material Interrogation Using Conductive Interfaces and Acoustic Wavefronts*, SIAM FR **21**, Philadelphia, 2002.

[BBu1] H. T. Banks and J. A. Burns, An abstract framework for approximate solutions to optimal control problems governed by hereditary systems, *International Conference on Differential Equations* (H. Antosiewicz, ed.), Academic Press (1975), 10–25.

[BBu2] H. T. Banks and J. A. Burns, Hereditary control problems: numerical methods based on averaging approximations, *SIAM Journal on Control and Optimization*, **16** (1978), 169–208.

[BCDST] H. T. Banks, F. Charles, M. Doumic-Jauffret, K. L. Sutton, and W. C. Thompson, Label structured cell proliferation models, CRSC-TR10-10, June, 2010; *Applied Mathematics Letters*, **23** (2010), 1412–1415.

[BCCW] H. T. Banks, D. Cioranescu, A. K. Criner, and W. P. Winfree, Modeling the flash-heat experiment on porous domains, CRSC-TR10-06, May, 2010; *Quarterly of Applied Mathematics*, S 0033-569X(2011)01230-8 (15 pages).

[BCCW2] H. T. Banks, D. Cioranescu, A. K. Criner, and W. P. Winfree, Parameter estimation for the heat equation on perforated domains, CRSC-TR11-06, April, 2011; *Journal of Inverse and Ill-posed Problems*, DOI.10.1515/JIIP.2011.051 (33 pages).

[BDK] H. T. Banks, P. L. Daniel, and P. M. Kareiva, Estimation techniques for transport equations, LCDS Technical Report #83-23, Brown Unversity, July, 1983; *Proceedings of the Conference on Mathematics in Biology and Medicine*, Lecture Notes in Biomathematics, Springer-Verlag (1985), 428–438.

[BDSS] H. T. Banks, M. Davidian, J. R. Samuels, Jr., and K. L. Sutton, An inverse problem statistical methodology summary, Center for Research in Scientific Computation Report, CRSC-TR08-01, January, 2008; Chapter 11 in *Mathematical and Statistical Estimation Approaches in Epidemiology*, (G. Chowell, M. Hyman, N. Hengartner, L. M. A. Bettencourt and C. Castillo-Chavez, eds.), Springer, Berlin (2009), pp. 249–302.

[BD] H. T. Banks and J. L. Davis, A comparison of approximation methods for the estimation of probability distributions on parameters, CRSC-TR05-38, 2005; *Applied Numerical Mathematics*, **57** (2007), 753–777.

[BDTR] H. T. Banks and J. L. Davis, Quantifying uncertainty in the estimation of probability distributions with confidence bands, CRSC-TR07-21, December, 2007; *Mathematical Biosciences and Engineering*, **5** (2008), 647–667.

[GRD-FP] H. T. Banks, J. L. Davis, S. L. Ernstberger, S. Hu, E. Artimovich, A. K. Dhar, and C. L. Browdy, A comparison of probabilistic and stochastic formulations in modeling growth uncertainty and variability, CRSC-TR08-03, February, 2008; *Journal of Biological Dynamics*, **3** (2009), 130–148.

[BDEHAD] H. T. Banks, J. L. Davis, S. L. Ernstberger, S. Hu, E. Artimovich, and A. K. Dhar, Experimental design and estimation of growth rate distributions in size-structured shrimp populations, CRSC-TR08-20, November, 2008; *Inverse Problems,* **25** (2009), 095003(28 pages), Sept.

[GRD-FP2] H. T. Banks, J. L. Davis, and S. Hu, A computational comparison of alternatives to including uncertainty in structured population models, CRSC-TR09-14, June, 2009; in *Three Decades of Progress in Systems and Control,* (X. Hu, et al., eds.) Springer, New York (2010), 19–33.

[BDEK] H. T. Banks, S. Dediu, S. L. Ernstberger, and F. Kappel, Generalized sensitivities and optimal experimental design, CRSC-TR08-12, September, 2008, Revised November, 2009; *Journal of Inverse and Ill-Posed Problems,* **18** (2010), 25–83.

[BDCI] H. T. Banks, S. Durso, M. G. Choi, and K. Ito, Nonlinear exothermic contributions to radio-frequency bonding of adhesives, CRSC-TR98-24, June 1998; *Nonlinear Analysis: Theory, Method and Applications: Series B,* **2** (2001), 257–386.

[BDGJ] H. T. Banks, S. R. Durso, M. A. Goodhart, and M. L. Joyner, On the radio-frequency inputs in dipolar heating of adhesives, CRSC-TR98-3, January, 1998; *Journal of Microwave Power and Electromagnetic Energy,* **33** (1998), 231–242.

[BFW88] H. T. Banks, R. Fabiano, and Y. Wang, Estimation of Boltzmann damping coefficients in beam models, LCDS/CCS Rep. 88-13, Brown University, July, 1988; *COMCON Conference on Stabilization of Flexible Structures,* (Montpellier, December, 1987), Optimization Software, Inc., New York (1988), 13–35.

[BFW89] H. T. Banks, R. Fabiano, and Y. Wang, Inverse problem techniques for beams with tip body and time hysteresis damping, ICASE Rep. #89-22 (April, 1989), NASA Langley Research Center, Hampton, VA; *Matematica Aplicada e Computacional,* **8** (1989), 101–118.

[BFWIC] H. T. Banks, R. Fabiano, Y. Wang, D. J. Inman, and H. Cudney, Spatial versus time hysteresis in damping mechanisms, *Proceedings of the 27th IEEE Conference on Decision and Control,* Austin, TX, December, 1988, 1674–1677.

[BF] H. T. Banks and B. G. Fitzpatrick, Estimation of growth rate distributions in size-structured population models, CAMS Tech. Rep. 90-2, University of Southern California, January, 1990; *Quarterly of Applied Mathematics*, **49** (1991), 215–235.

[BFPZ] H. T. Banks, B. G. Fitzpatrick, L. K. Potter, and Y. Zhang, Estimation of probability distributions for individual parameters using aggregate population data, CRSC-TR98-6, January, 1998; In *Stochastic Analysis, Control, Optimization and Applications*, (W. McEneaney, G. Yin, and Q. Zhang, eds.), Birkhaüser, Boston, 1989.

[BG1] H. T. Banks and N. L. Gibson, Well-posedness in Maxwell systems with distributions of polarization relaxation parameters, CRSC-TR04-01, January, 2004; *Applied Mathematics Letters*, **18** (2005), 423–430.

[BG2] H. T. Banks and N. L. Gibson, Electromagnetic inverse problems involving distributions of dielectric mechanisms and parameters, CRSC-TR05-29, August, 2005; *Quarterly of Applied Mathematics*, **64** (2006), 749–795.

[BGIT] H. T. Banks, S. L. Grove, K. Ito, and J. A. Toivanen, Static two-player evasion-interrogation games with uncertainty, CRSC-TR06-16, June, 2006; *Journal of Computational and Applied Mathematics*, **25** (2006), 289–306.

[BHK] H. T. Banks, K. J. Holm, and F. Kappel, Comparison of optimal design methods in inverse problems, *Inverse Problems*, **27** (2011), 075002.

[BHK-TR] H.T. Banks, K. J. Holm, and F. Kappel, Comparison of optimal design methods in inverse problems, CRSC Technical Report, CRSC-TR10-11, July, 2010.

[BHR] H. T. Banks, K. J. Holm, and D. Robbins, Standard error computations for uncertainty quantification in inverse problems: asymptotic theory vs. bootstrapping, CRSC-TR09-13, June, 2009; Revised, August 2009; *Mathematical and Computer Modeling*, **52** (2010), 1610–1625.

[BHu] H. T. Banks and S. Hu, Nonlinear stochastic Markov processes and modeling uncertainty in populations, CRSC-TR11-02, January, 2011; *Mathematical Biosciences and Engineering*, **9** (2012), 1–25.

[BI86] H. T. Banks and D. W. Iles, On compactness of admissible parameter sets: convergence and stability in inverse problems for distributed parameter systems, ICASE Report #86-38, NASA Langley Research Center, Hampton VA 1986; *Proceedings of the Conference on Control Systems Governed by PDEs*, February, 1986, Gainesville, FL, *Springer Lecture Notes in Control & Information Science*, **97** (1987), 130–142.

[BaIn] H. T. Banks and D. J. Inman, On damping mechanisms in beams, ICASE Rep. #89-64 August, 1989, NASA LaRC, Hampton, VA; *ASME Journal of Applied Mechanics*, **58**, (1991), 716–723.

[BI] H. T. Banks and K. Ito, A unified framework for approximation and inverse problems for distributed parameter systems, ICASE/NASA Report No. 88-12, January, 1988, NASA LaRC, Hampton, VA; *Control: Theory and Advanced Technology*, **4** (1988), 73–90.

[BI2] H. T. Banks and K. Ito, Approximation in LQR problems for infinite dimensional systems with unbounded input operators, CRSC-TR94-22, November, 1994; electronic publication *Journal of Mathematical Systems, Estimation and Control*, **7** (1997), 1–34; summarized in print *Journal of Mathematical Systems, Estimation and Control*, **7** (1997), 119–122.

[BIKT] H. T. Banks, K. Ito, G. M. Kepler, and J. A. Toivanen, Material surface design to counter electromagnetic interrogation of targets, CRSC-TR04-37, November, 2004; *SIAM Journal on Applied Mathematics*, **66** (2006), 1027–1049.

[BIT] H. T. Banks, K. Ito and J. A. Toivanen, Determination of interrogating frequencies to maximize electromagnetic backscatter from objects with material coatings, CRSC-TR05-30, August, 2005; *Communications in Computational Physics*, **1** (2006), 357–377.

[BaItWa] H. T. Banks, K. Ito, and C. Wang, Exponentially stable approximations of weakly damped wave equations, in *Distributed Parameter Systems Control and Applications*, (F. Kappel et al., eds.),
ISNM Vol. 100, Birkhäuser, Boston, 1991, 1–33.

[BKap1] H. T. Banks and F. Kappel, Spline approximations for functional differential equations, *Journal of Differential Equations*, **34**
(1979), 496–522.

[BKap] H. T. Banks and F. Kappel, Transformation semigroups and L^1
approximation for size structured population models, LCDS/CCS
Rep. 88-20, Brown University, July, 1988; *Semigroup Forum*, **38**
(1989), 141–155.

[BKW1] H. T. Banks, F. Kappel, and C. Wang, Weak solutions
and differentiability for size structured population models, CAMS
Tech. Rep. 90-11, August, 1990, University of Southern California; in *Distributed Parameter Systems Control and Applications*,
Birkhäuser, Boston, *International Series of Numerical Mathematics*, Vol. **100** (1991), 35–50.

[BKW2] H. T. Banks, F. Kappel, and C. Wang, A semigroup formulation of nonlinear size-structured distributed rate population model, CRSC-TR94-4, March 1994; in *Control and Estimation of Distributed Parameter Systems: Nonlinear Phenomena*,
Birkhäuser, Boston, ISNM, **118** (1994), 1–19.

[BKa] H. T. Banks and P. Kareiva, Parameter estimation techniques
for transport equations with application to population dispersal
and tissue bulk flow models, LCDS Report #82-13, Brown University, July 1982; *Journal of Mathematical Biology*, **17** (1983),
253–273.

[BKL] H. T. Banks, P. M. Kareiva and P. K. Lamm, Modeling insect
dispersal and estimating parameters when mark-release techniques
may cause initial disturbances, *Journal of Mathematical Biology*,
22 (1985), 259–277.

[BKZ] H. T. Banks, P. M. Kareiva, and L. Zia, Analyzing field studies of insect dispersal using two-dimensional transport equations,
LCDS/CCS Rep. 86-48, Brown University, November, 1986; *Environmental Entomology*, **17** (1988), 815–820.

[BKT1] H. T. Banks, Z. R. Kenz, and W. C. Thompson, Computational comparison of methods for the nonparametric estimation of probability measures, to appear.

[BKT2] H. T. Banks, Z. R. Kenz, and W. C. Thompson, A survey of selected techniques in inverse problem nonparametric probability distribution estimation, to appear.

[BKo1] H. T. Banks and F. Kojima, Boundary shape identification problems in two dimensional domains related to thermal testing of materials, LCDS/CCS Rep. 88-6, Brown University, April, 1988; *Quarterly of Applied Mathematics*, **47** (1989), 273–293.

[BKo2] H. T. Banks and F. Kojima, Boundary identification for 2-D parabolic systems arising in thermal testing materials, *Proceedings of 27th IEEE Confernece on Decision and Control*, Austin, TX, December, 1988, 1678–1683.

[BKoW] H. T. Banks, F. Kojima, and W. P. Winfree, Boundary estimation problems arising in thermal tomography, CAMS Tech. Rep. 89-6, University of Southern California; *Inverse Problems*, **6** (1990), 897–921.

[BKregulator] H. T. Banks and K. Kunisch, The linear regulator problem for parabolic systems, *SIAM Journal on Control and Optimization*, **22** (1984), 684–698.

[BK] H. T. Banks and K. Kunisch, *Estimation Techniques for Distributed Parameter Systems*. Birkhauser, Boston, 1989.

[BKcontrol] H. T. Banks and K. Kunisch, An approximation theory for nonlinear partial differential equations with applications to identification and control, LCDS Technical Report #81-7, Brown University, April 1981; *SIAM Journal on Control and Optimization*, **20** (1982), 815–849.

[BanksKurdila] H. T. Banks and A. J. Kurdila, Hysteretic control influence operators representing smart material actuators: Identification and approximation, *Proceedings of IEEE Conference on Decision and Control*, Kobe, Japan, December, 1996, 3711–3716.

[BKurW1] H. T. Banks, A. J. Kurdila, and G. Webb, Identification of hysteretic control influence operators representing smart actuators, Part I: Formulation, *Mathematical Problems in Engineering*, **3** (1997), 287–328.

[BKurW2] H. T. Banks, A. J. Kurdila, and G. Webb, Identification of hysteretic control influence operators representing smart actuators, Part II: Convergent approximations, *Journal of Intelligent Material Systems and Structures*, **8** (1997), 536–550.

[BaPed] H. T. Banks and M. Pedersen, Well-posedness of inverse problems for systems with time dependent parameters, CRSC-TR08-10, August, 2008; *Arabian Journal of Science and Engineering: Mathematics,* **1** (2009), 39–58.

[BPi] H. T. Banks and G. A. Pinter, A probabilistic multiscale approach to hysteresis in shear wave propagation in biotissue, CRSC-TR04-03, January, 2004; *SIAM Journal on Multiscale Modeling and Simulation*, **3** (2005), 395–412.

[BPo] H. T. Banks and L. K. Potter, Probabilistic methods for addressing uncertainty and variability in biological models: application to a toxicokinetic model, CRSC-TR02-27, September, 2002; *Mathematical Biosciences*, **192** (2004), 193–225.

[BR1] H. T. Banks and I. G. Rosen, Numerical schemes for the estimation of functional parameters in distributed models for mixing mechanisms in lake and sea sediment cores, LCDS Technical Report #85-27, October, 1985; *Inverse Problems*, **3** (1987), 1–23.

[BR2] H. T. Banks and I. G. Rosen, Approximation methods for the solution of inverse problems in lake and sea sediment core analysis, *Proceedings of 24th IEEE Conference on Decision and Control,* December, 1985, Ft. Lauderdale, FL.

[BRI] H. T. Banks, I. G. Rosen, and K. Ito, A spline-based technique for computing Riccati operators and feedback controls in regulator problems for delay systems, *SIAM Journal on Scientific and Statistical Computing*, **5** (1984), 830–855.

[BRS] H. T. Banks, K. L. Rehm, and K. L. Sutton, Dynamic social network models incorporating stochasticity and delays, CRSC-TR09-

11, May, 2009; *Quarterly of Applied Mathematics*, **68** (2010), 783–802.

[BSW] H. T. Banks, R. C. Smith, and Y. Wang, *Smart Material Structures: Modeling, Estimation and Control*, Masson/John Wiley, Paris/Chichester, 1996.

[BSTBRSM] H. T. Banks, K. L. Sutton, W. C. Thompson, G. Bocharov, D. Roose, T. Schenkel, and A. Meyerhans, Estimation of cell proliferation dynamics using CFSE data, CRSC-TR09-17, August, 2009; *Bulletin of Mathematical Biology*, **73** (2011), 116–150.

[BST2] H. T. Banks, K. L. Sutton, W. C. Thompson, G. Bocharov, M. Doumic, T. Schenkel, J. Argilaguet, S. Giest, C. Peligero, and A. Meyerhans, A new model for the estimation of cell proliferation dynamics using CFSE data, CRSC-TR11-05, March, 2011; *Journal of Immunological Methods*, **373** (2011), 143–160; DOI:10.1016/j.jim.2011.08.014.

[BT] H. T. Banks and H. T. Tran, *Mathematical and Experimental Modeling of Physical and Biological Processes*, CRC Press, Boca Raton, FL, 2009.

[BWIC] H. T. Banks, Y. Wang, D. J. Inman, and H. Cudney, Parameter identification techniques for the estimation of damping in flexible structure experiments, *Proceedings of 26th IEEE Conference on Decision and Control*, December, 1987, Los Angeles, 1392–1395.

[Barbu] V. Barbu, *Nonlinear Semigroups and Differential Equations in Banach Spaces*, Noordhoff, Leyden, 1976.

[Basar] T. Basar and G. J. Olsder, *Dynamic Noncooperative Game Theory*, SIAM Publ., Philadelphia, PA, 2nd edition, 1999.

[BA] G. Bell and E. Anderson, Cell growth and division I. A mathematical model with applications to cell volume distributions in mammalian suspension cultures, *Biophysical Journal*, **7** (1967), 329–351.

[BellmanCooke] R. Bellman and K. L. Cooke, *Differential-Difference Equations*, Vol. **6**, Mathematics in Science and Engineering, Academic Press, New York, 1963.

[BW] M. P. F. Berger and W. K. Wong (eds.), *Applied Optimal Designs*, John Wiley & Sons, Chichester, UK, 2005.

[Berkovitz] L. D. Berkovitz, *Optimal Control Theory*, Springer-Verlag, New York, NY, 1974.

[BernHyland] D. S. Bernstein and D. C. Hyland, The optimal projection equations for finite-dimensional fixed-order dynamic compensation of infinite-dimensional systems, *SIAM Journal on Control and Optimization*, **24** (1986), 122–151.

[KLB] K. L. Bihari, *Analysis of Thermal Conductivity in Composite Adhesives*, PhD Thesis, North Carolina State University, Raleigh, NC, 2001.

[B] P. Billingsley, *Convergence of Probability Measures*, Wiley, New York, 1968.

[BR] G. Birkhoff and G. C. Rota, *Ordinary Differential Equations*, Ginn and Co., Boston, 1962.

[Bortz02] D. M. Bortz, *Modeling, Analysis, and Estimation of an in vitro HIV Infection Using Functional Differential Equations*, Ph.D. Thesis, N. C. State University, Raleigh, NC, 2002 (http://www.lib.ncsu.edu).

[Bortz, et al.] D. M. Bortz, R. Guy, J. Hood, K. Kirkpatrick, V. Nguyen, and V. Shimanovich, Modeling HIV infection dynamics using delay equations, in: P. A. Gremaud, Z. Li, R. C. Smith, H. T. Tran (eds.), *Proceedings of the 2000 Industrial Mathematics Modeling Workshop for Graduate Students*, 2000, pp. 48–58, CRSC-TR00-24, October, 2000.

[BroSpre] M. Brokate and J. Sprekels, *Hysteresis and Phase Transitions*, Applied Math Sciences, Vol. **121**, Springer-Verlag, New York, 1996.

[ButBer] P. L. Butzer and H. Berens, *Semi-Groups of Operators and Approximation*, Springer-Verlag, New York, 1967.

[CJ] H. S. Carslaw and J. C. Jaeger, *Conduction of Heat in Solids*, Oxford University Press, 1959.

[Cheng] D. K. Cheng, *Field and Wave Electromagnetics*, Addison Wesley, Reading, MA, 1983.

[CK1] D. Colton and R. Kress, *Integral Equation Methods in Scattering Theory*, John Wiley & Sons, New York, 1983.

[CK2] D. Colton and R. Kress, *Inverse Acoustic and Electromagnetic Scattering Theory*, Springer-Verlag, New York, 1992.

[CH] R. Courant and D. Hilbert, *Methods of Mathematical Physics*, Vol. 1, Wiley Interscience, New York, 1953.

[CoxMiller] D. R. Cox and H. D. Miller, *The Theory of Stochastic Processes*, Chapman and Hall, London, 1965.

[Curtain84] R. F. Curtain, Finite dimensional compensators for parabolic distributed systems with unbounded control and observation, *SIAM Journal on Control and Optimization*, **22** (1984), 255–276.

[CurtainSalamon] R. F. Curtain and D. Salamon, Finite dimensional compensators for infinite dimensional systems with unbounded input operators, *SIAM Journal on Control and Optimization*, **24** (1986), 797–816.

[Cushing] J. M. Cushing, *Integrodifferential Equations and Delay Models in Population Dynamics*, No. **20** in Lecture Notes in Biomathematics, Springer-Verlag, New York, 1977.

[DaP] G. Da Prato, Synthesis of optimal control for an infinite dimensional periodic problem, *SIAM Journal on Control and Optimization*, **25** (1987), 706–714.

[DaPrato] G. Da Prato, et al., eds., *Functional Analytic Methods for Evolution Equations*, Springer-Verlag, Berlin-Heidelberg, 2004.

[DTB] J. David, H. T. Tran, and H. T. Banks, Receding horizon control of HIV, CRSC-TR09-21, N.C. State University, December, 2009; *Optimal Control, Applications and Methods*, **32** (2011), 681–699.

[DavidianGallant] M. Davidian and A. R. Gallant, Smooth nonparametric maximum likelihood estimation for population pharmacokinetics, with application to quinidine, *Journal of Pharmacokinetics and Biopharmaceutics*, **20** (1992), 529–556.

[DavidianGallant2] M. Davidian and A. R. Gallant, The nonlinear mixed effects model with a smooth random effects density, *Biometrika*, **80** (1993), 475–488.

[DG] M. Davidian and D. Giltinan, *Nonlinear Models for Repeated Measurement Data*, Chapman & Hall, London, 1998.

[D] J. W. Dettman, *Mathematical Methods in Physics and Engineering*, McGraw-Hill, New York, 1962.

[DBW] F. Dexter, H. T. Banks, and T. Webb III, Modeling Holocene changes of the location and abundance of Beech populations in eastern North America, LCDS Technical Report #86-36, Brown University, September 1986; *Review of Palaeobotany and Palynology*, **50** (1987), 273–292.

[Diekmannetal] O. Diekmann, S. A. van Gils, S. M. V. Lunel, and H. O. Walther, *Delay Equations: Functional-, Complex-, and Nonlinear Analysis*, No. **110** in Applied Mathematical Sciences, Springer-Verlag, New York, 1995.

[Driver] R. D. Driver, *Ordinary and Delay Differential Equations*, No. **20** in Applied Mathematical Sciences, Springer-Verlag, New York, 1977.

[DS] N. Dunford and J. T. Schwartz, *Linear Operators*, Vol. I, John Wiley & Sons, New York, 1966.

[E] R. E. Edwards, *Functional Analysis: Theory and Applications*, Holt, Rinehart and Winston, New York, 1965.

[Elliot] R. S. Elliot, *Electromagnetics: History, Theory, and Applications*, IEEE Press, New York, 1993.

[Ev] L. C. Evans, *Partial Differential Equations*, Graduate Studies in Mathematics, Vol. **19**, Amer. Math. Soc., Providence, 2002.

[Ewing] G. M. Ewing, *Calculus of Variations with Applications*, W.W. Norton, New York, 1969.

[Fed] V. V. Fedorov, *Theory of Optimal Experiments*, Academic Press, New York, 1972.

[FedHac] V. V. Fedorov and P. Hackel, *Model-Oriented Design of Experiments,* Springer-Verlag, New York, 1997.

[Filippov62] A. F. Filippov, On certain questions in the theory of optimal control, *SIAM Journal on Control,* **1** (1962), 76–84.

[FrischHolme] F. Frisch and H. Holme, The characteristic solution of a mixed difference and differential equation occurring in economic dynamics, *Econometrica,* **3** (1935), 225–239.

[GallantNychka] A. R. Gallant and D. Nychka, Semi-nonparametric maximum likelihood estimation, *Econometrica,* **55** (1987), 363–390.

[Gard] T. C. Gard, *Introduction to Stochastic Differential Equations,* Marcel Dekker, New York, 1988.

[GA90] J. S. Gibson and A. Adamian, Approximation theory for linear-quadratic-Gaussian optimal control of flexible structures, *SIAM Journal on Control and Optimization,* **29** (1991), 1–37.

[GP] C. Goffman and G. Pedrick, *First Course in Functional Analysis,* Prentice-Hall, Englewood Cliffs, NJ, 1965.

[Goldstein] G. R. Goldstein et al., eds., *Evolution Equations: Proceedings in Honor of J.A. Goldstein's 60th Birthday,* Marcel Dekker, New York, 2003.

[Goreckietal] H. Górecki, S. Fuksa, P. Grabowski, and A. Korytowski, *Analysis and Synthesis of Time Delay Systems,* John Wiley & Sons, New York, 1989.

[Grossman,etal] Z. Grossman, M. Feinberg, V. Kuznetsov, D. Dimitrov, and W. Paul, HIV infection: how effective is drug combination treatment? *Immunology Today,* **19** (1998), 528–532.

[JKH1] J. K. Hale, *Functional Differential Equations,* Springer-Verlag, New York, 1971.

[JKH2] J. K. Hale, *Theory of Functional Differential Equations,* Springer-Verlag, New York, 1977.

[JKH3] J. K. Hale and S. M. Verduyn Lunel, *Introduction to Functional Differential Equations,* Springer-Verlag, New York, 1993.

[JKH4] J. K. Hale, *Ordinary Differential Equations*, Wiley, New York, 1969.

[Hal] P. R. Halmos, *Measure Theory*, Springer-Verlag, Berlin-New York, 1974

[HKNT] E. Heikkola, Y. A. Kuznetsov, P. Neittaanmäki, and J. Toivanen, Fictitious domain methods for the numerical solution of two-dimensional scattering problems, *Journal of Computational Physics*, **145** (1998), 89–109.

[HRT] E. Heikkola, T. Rossi, and J. Toivanen, Fast direct solution of the Helmholtz equation with a perfectly matched layer or an absorbing boundary condition, *International Journal for Numerical Methods in Engineering*, **57** (2003), 2007–2025.

[HeSt] E. Hewitt and K. R. Stromberg, *Real and Abstract Analysis*, Springer-Verlag, New York, 1965.

[HP] E. Hille and R. S. Phillips, *Functional Analysis and Semigroups*, AMS Colloquium Publications, Providence, RI, Vol. **31**, 1957.

[HolteEmerman] S. Holte and M. Emerman, A competition model for viral inhibition of host cell proliferation, *Mathematical Biosciences*, **166** (2000), 69–84.

[Hutch] G. E. Hutchinson, Circular causal systems in ecology, *Annals of the New York Academy of Sciences*, **50** (1948), 221–246.

[H] P. Huber, *Robust Statistics*, John Wiley & Sons, New York, 2004.

[IncropDewitt] F. Incropera and D. Dewitt, *Fundamentals of Heat and Mass Transfer*, John Wiley & Sons, New York, 1990.

[Ito90] K. Ito, Finite-dimensional compensators for infinite-dimensional systems via Galerkin-type approximations, *SIAM Journal on Control and Optimization*, **28** (1990), 1251–1269.

[IT] K. Ito and H. T. Tran, Linear quadratic optimal control problem for linear systems with unbounded input and output operators: numerical approximations, *Proceedings of the 4th International Conference on Control of Distributed Systems*, (F. Kappel et al., eds.), ISNM Vol. 91, Birkhäuser, Boston, 1989, pp. 171-195.

[Jackson] J. D. Jackson, *Classical Electrodynamics*, John Wiley & Sons, New York, 1975.

[Jensen] A. Jensen, An elucidation of Erlang's statistical works through the theory of stochastic processes, in: E. Brockmeyer, H. L. Halstrøm, and A. Jensen (eds.), *The Life and Works of A. K. Erlang*, The Copenhagen Telephone Company, Copenhagen, 1948, pp. 23–100.

[Kappel82] F. Kappel, An approximation scheme for delay equations, in: V. Lakshmikantham (ed.), *Nonlinear Phenomena in Mathematical Sciences*, Academic Press, New York, 1982, pp. 585–595.

[KapSal87] F. Kappel and D. Salamon, Spline approximation for retarded systems and the Riccati equation, *SIAM Journal on Control and Optimization*, **25** (1987), 1082–1117.

[KapSal89] F. Kappel and D. Salamon, On the stability properties of spline approximations for retarded systems, *SIAM Journal on Control and Optimization*, **27** (1989), 407–431.

[KapSch] F. Kappel and W. Schappacher, Autonomous nonlinear functional differential equations and averaging approximations, *Journal of Nonlinear Analysis*, **2** (1978), 391–422.

[K] T. Kato, *Perturbations Theory for Linear Operators*, Springer-Verlag, New York, 1966.

[Kot] M. Kot, *Elements of Mathematical Ecology*, Cambridge University Press, Cambridge, 2001.

[KP] M. A. Krasnoselskii and A. V. Pokrovskii, *Systems with Hysteresis*, Springer-Verlag, New York, 1989.

[KS] C. Kravis and J. H. Seinfield, Identification of parameters in distributed parameter systems by regularization, *SIAM Journal on Control and Optimization*, **23** (1985), 217–241.

[Kreĭn] S. G. Kreĭn, *Linear Differential Equations in Banach Spaces*, American Mathematical Society, Providence, RI, 1972.

[Kress] R. Kress, *Linear Integral Equations*, Springer-Verlag, New York, 1989.

[Kreyszig] E. Kreyszig, *Introductory Functional Analysis with Applications*, Wiley Classics Edition, John Wiley & Sons, New York, 1989.

[Kuang] Y. Kuang, *Delay Differential Equations with Applications in Population Dynamics*, Vol. **191** of Mathematics in Science and Engineering, Academic Press, New York, 1993.

[Las92] I. Lasiecka, Galerkin approximations of infinite-dimensional compensators for flexible structures with unbounded control action, *Acta Applicandae Mathematicae*, **28** (1992), 101–133.

[Las95] I. Lasiecka, Finite element approximations of compensator design for analytic generators with fully unbounded controls/observations, *SIAM Journal on Control and Optimization*, **33** (1995), 67–88.

[LLP95] I. Lasiecka, D. Lukes, and L. Pandolfi, Input dynamics and nonstandard Riccati equations with applications to boundary control of damped wave and plate equations, *Journal of Optimization Theory and Applications*, **84** (1995), 549–574.

[Lindsay] B. G. Lindsay, The geometry of mixture likelihoods: a general theory, *Annals of Statistics*, **11** (1983), 86–94.

[Lindsay2] B. G. Lindsay, *Mixture Models: Theory, Geometry, and Applications*, NSF-CBMS Regional Conference Series in Probability and Statistics, Institute of Mathematical Statistics, Haywood, CA, Vol. **5**, 1995.

[Li] J. L. Lions, *Optimal Control of Systems Governed by Partial Differential Equations*, Springer-Verlag, New York, 1971.

[LiMag] J. L. Lions and E. Magenes, *Non-Homogeneous Boundary Value Problems and Applications*, Vol.I-III, Springer-Verlag, Berlin, 1972.

[LiuSob] L. Liusternik and V. Sobolev, *Elements of Functional Analysis*, Frederick Unger Publishing, New York, 1961.

[Lloyd] A. L. Lloyd, The dependence of viral parameter estimates on the assumed viral load life cycle: limitations of studies of viral load data, *Proceedings of the Royal Society of London Series B*, **268** (2001), 847–854.

[Lloyd2] A. L. Lloyd, Destabilization of epidemic models with the inclusion of realistic distributions of infectious periods, *Proceedings of the Royal Society of London Series B*, **268** (2001), 985–993.

[Luen69] D. G. Luenberger, *Optimization by Vector Space Methods*, John Wiley & Sons, New York, 1969.

[Mallet] A. Mallet, A maximum likelihood estimation method for random coefficient regression models, *Biometrika*, **73** (1986), 645–656.

[Mallet2] A. Mallet, F. Mentré, J. L. Steimer, and F. Lokiec, Nonparametric maximum likelihood estimation for population pharmacokinetics with application to cyclosporine, *Journal of Pharmacokinetics and Biopharmaceutics*, **16** (1988), 311–327.

[Martin] R. H. Martin, Jr., *Nonlinear Operators and Differential Equations in Banach Spaces*, John Wiley & Sons, New York, 1976.

[May] R. M. May, *Stability and Complexity in Model Ecosystems*, Princeton Landmarks in Biology, Princeton University Press, Princeton, NJ, 2001.

[Ma] I. D. Mayergoyz, *Mathematical Models of Hysteresis*, Springer-Verlag, New York, 1991.

[McS] E. J. McShane, *Integration*, Princeton University Press, Princeton, NJ, 1944.

[McShane40-1] E. J. McShane, Necessary conditions in generalized-curve problems of the calculus of variations, *Duke Mathematical Journal*, **7** (1940), 1–27.

[McShane40-2] E. J. McShane, Generalized curves, *Duke Mathematical Journal*, **6** (1940), 513–536.

[McShane67] E. J. McShane, Relaxed controls and variational problems, *SIAM Journal on Control*, **5** (1967), 438–485.

[MetzD] J. A. J. Metz and E. O. Diekmann, *The Dynamics of Physiologically Structured Populations*, Lecture Notes in Biomathematics, Vol. **68**, Springer, Heidelberg, 1986.

[Minorsky] N. Minorsky, Self-excited oscillations in a dynamical system possessing retarded actions, *Journal of Applied Mechanics*, **9** (1942), 65–71.

[MitSulNeumPerel] J. E. Mittler, B. Sulzer, A. U. Neumann, and A. S. Perelson, Influence of delayed viral production on viral dynamics in HIV-1 infected patients, *Mathematical Biosciences*, **152** (1998), 143–163.

[Murray] J. D. Murray, *Mathematical Biology*, Vol. **19** in Biomathematics, Springer-Verlag, New York, 1989.

[NS] A. W. Naylor and G. R. Sell, *Linear Operator Theory in Engineering and Science*, Springer Verlag, New York, 1982.

[NelsonMittlerPerelson] P. W. Nelson, J. E. Mittler, and A. S. Perelson, Effect of drug efficacy and the eclipse phase of the viral life cycle on estimates of HIV viral dynamic parameters, *Journal of Acquired Immune Deficiency Syndromes*, **26** (2001), 405–412.

[NelsonMurrayPerelson] P. W. Nelson, J. D. Murray, and A. S. Perelson, A model of HIV-1 pathogenesis that includes an intracellular delay, *Mathematical Biosciences*, **163** (2000) 201–215.

[NelsonPerelson] P. W. Nelson and A. S. Perelson, Mathematical analysis of delay differential equation models of HIV-1 infection, *Mathematical Biosciences* **179** (2002), 73–94.

[NowakMay] M. A. Nowak and R. M. May, *Virus Dynamics: Mathematical Principles of Immunology and Virology*, Oxford University Press, New York, 2000.

[Okubo] A. Okubo, *Diffusion and Ecological Problems: Mathematical Models*, Biomathematics, **10**, Springer-Verlag, Berlin, 1980.

[Pa] A. Pazy, *Semigroups of Linear Operators and Applications to Partial Differential Equations*, Springer-Verlag, New York, 1983.

[Ped] M. Pedersen, *Functional Analysis in Applied Mathematics and Engineering*, Chapman & Hall/CRC Press, Boca Raton, 2000.

[Po] L. K. Potter, *Physiologically Based Pharmacokinetic Models for the Systemic Transport of Trichloroethylene*, Ph.D. Thesis, N.C. State University, August, 2001; (http://www.lib.ncsu.edu).

[Preisach35] F. Preisach, Über die magnetische Nachwirkung, *Zeitschrift für Physik*, **94** (1935), 277–302.

[PS] A. J. Pritchard and D. Salamon, The linear quadratic control problem for infinite dimensional systems with unbounded input and output operators, *SIAM Journal on Control and Optimization*, **25** (1987), 121–144.

[P] Yu. V. Prohorov, Convergence of random processes and limit theorems in probability theory, *Theory of Probability and Its Applications*, **1** (1956), 157–214.

[RM] R. D. Richtmyer and K. W. Morton, *Finite Differences Methods for Initial Value Problems*, 2nd ed., Wiley Interscience, New York, 1967.

[RN] F. Riesz and B. S. Nagy, *Functional Analysis*, Ungar Press, 1955.

[RMC] J. R. Reitz, R. W. Christy, and F. J. Milford, *Foundations of Electromagnetic Theory*, Addison Wesley, Reading, MA, 1992.

[RogelWuEmerman] M. E. Rogel, L. I. Wu, and M. Emerman, The human immunodeficiency virus type 1 vpr gene prevents cell proliferation during chronic infection, *Journal of Virology*, **69** (1995), 882–888.

[Royden] H. L. Royden, *Real Analysis*, 2nd ed., Macmillan, New York, 1968.

[Ru1] W. Rudin, *Principles of Mathematical Analysis*, McGraw-Hill, New York, 1953.

[Ru2] W. Rudin, *Real and Complex Analysis*, McGraw-Hill, New York, 1966.

[DLR] D. L. Russell, On mathematical models for the elastic beam with frequency proportional damping, in *Control and Estimation of Distributed Parameter Systems*, (Frontiers in Applied Mathematics **11**), Philadelphia, SIAM, 1992, 125–169.

[Sagan] H. Sagan, *Introduction to the Calculus of Variations*, McGraw-Hill, New York, 1969.

[Schultz] M. H. Schultz, *Spline Analysis*, Prentice-Hall, Englewood Cliffs, NJ, 1973.

[SchuVarg] M. H. Schultz and R. S. Varga, L-splines, *Numerische Mathematik*, **10** (1967), 345–369.

[Sch83] J. M. Schumacher, A direct approach to compensator design for distributed parameter systems, *SIAM Journal on Control and Optimization*, **21** (1983), 823–836.

[Schumitzky] A. Schumitzky, Nonparametric EM algorithms for estimating prior distributions, *Applied Mathematics and Computation*, **45** (1991), 141–157.

[SeWi] G. A. F. Seber and C. J. Wild, *Nonlinear Regression,* John Wiley & Sons, New York, 1989.

[Sh] R. E. Showalter, *Hilbert Space Methods for Partial Differential Equations*, Pitman, London, 1977.

[SS] J. Sinko and W. Streifer, A new model for age-size structure of a population, *Ecology,* **48** (1967), 910–918.

[Smir] V. I. Smirnov, *Integration and Functional Analysis*, Vol. 5 of A Course of Higher Mathematics, Addison-Wesley, Reading, MA, 1964.

[RCS] R. C. Smith, *Smart Material Systems: Model Development*, Frontiers in Applied Math, **FR32**, SIAM, Philadelphia, 2005.

[S] I. Stakgold, *Boundary Value Problems of Mathematical Physics*, Vol. 1, Macmillan, New York, 1967.

[StrangFix] G. Strang and G. J. Fix, *An Analysis of the Finite Element Method*, Prentice-Hall, Englewood Cliffs, NJ, 1973.

[T] H. Tanabe, *Equations of Evolution*, Pitman, London, 1979.

[Vis] A. Visintin, *Differential Models of Hystersis*, Springer-Verlag, New York, 1994.

[Vog] C. R. Vogel, *Computational Methods for Inverse Problems*, Frontiers in Applied Math, **FR23**, SIAM, Philadelphia, 2002.

[VN] J. von Neumann, Zur theorie der gesellschaftsspiele, *Mathematische Annalen,* **100** (1928), 295–320.

[VNM] J. von Neumann and O. Morgenstern, *Theory of Games and Economic Behavior*, Princeton University Press, Princeton, NJ, 1944.

[Wahba1] G. Wahba, Bayesian "confidence intervals" for the cross-validated smoothing spline, *Journal of the Royal Statistical Society B,* **45** (1983), 133–150.

[Wahba2] G. Wahba, (Smoothing) splines in nonparametric regression, Dept of Statistics-TR 1024, Univ. Wisconsin, September, 2000; in *Encyclopedia of Environmetrics,* (A. El-Shaarawi and W. Piegorsch, eds.), Wiley, Vol. **4,** pp. 2099–2112, 2001.

[Warga62] J. Warga, Relaxed variational problems, *Journal of Mathematical Analysis and Applications,* **4** (1962), 111–128.

[Warga67] J. Warga, Functions of relaxed controls, *SIAM Journal on Control,* **5** (1967), 628–641.

[Warga72] J. Warga, *Optimal Control of Differential and Functional Equations,* Academic Press, New York, 1972.

[Webb] G. F. Webb, Functional differential equations and nonlinear semigroups in L_p spaces, *Journal of Differential Equations,* **20,** (1976), 71–89.

[W] J. Wloka, *Partial Differential Equations,* Cambridge University Press, Cambridge, 1987.

[Wright] E. M. Wright, A non-linear difference-differential equation, *Journal für die Reine Angewandte Mathematik,* **494** (1955), 66–87.

[YY] Y. S. Yoon and W. W. G. Yeh, Parameter identification in an inhomogeneous medium with the finite element method, *Society of Petroleum Engineers Journal,* **16** (1976), 217–226.

[Y] K. Yosida, *Functional Analysis,* 3rd ed., Springer-Verlag, Berlin, 1971.

[Young37] L. C. Young, Generalized curves and the existence of an attained absolute minimum in the calculus of variations, *Comptes Rendus Societe Sciences et Lettres,* Varsovie, Cl. III, **30** (1937), 212–234.

[Young38] L. C. Young, Necessary conditions in the calculus of variations, *Acta Mathematica,* **69** (1938), 239–258.

Index

Milton Keynes UK
Ingram Content Group UK Ltd.
UKHW031146141024
449569UK00024B/1038